The Cambridge Manuals of Science and
Literature

HEREDITY

HEREDITY

IN THE LIGHT OF RECENT RESEARCH

BY THE LATE

L. DONCASTER,
Sc.D., F.R.S.

Fellow of King's College, Cambridge
Derby Professor of Zoology,
Liverpool University

Cambridge :
at the University Press

1921

CAMBRIDGE UNIVERSITY PRESS
Cambridge, New York, Melbourne, Madrid, Cape Town,
Singapore, São Paulo, Delhi, Tokyo, Mexico City

Cambridge University Press
The Edinburgh Building, Cambridge CB2 8RU, UK

Published in the United States of America by
Cambridge University Press, New York

www.cambridge.org
Information on this title: www.cambridge.org/9781107401914

First edition 1910
Reprinted 1911
Second edition 1912
Third edition 1921
First paperback edition 2011

A catalogue record for this publication is available from the British Library

ISBN 978-1-107-40191-4 Paperback

*With the exception of the coat of arms
at the foot, the design on the title page is a
reproduction of one used by the earliest known
Cambridge printer, John Siberch, 1521*

PREFACE

IN a book of the size to which the Cambridge Manuals of Science and Literature are limited, it is plainly impossible to treat in detail every aspect of a subject like Heredity. One of the chief difficulties, therefore, which I have encountered in preparing this little book has been to decide what to leave out. To some it will doubtless seem that parts of the subject have been treated too fully, and other important branches omitted or barely mentioned, but my aim has been to give the reader a sketch of the most important lines in which recent advances have been made. There are many excellent works dealing with the older theories—and in this subject age is measured by very few years,—but our knowledge has increased so greatly and is still progressing so quickly that books become out of date almost as soon as they are published. My attempt, then, has been to deal chiefly with the quite modern developments of the subject, and in order that the reader who is not very familiar with the matter may feel he is on fairly sure ground, and not confuse fact with speculation, I have tried to avoid purely speculative questions in the body of the book, and have devoted a few pages to one of the most interesting of these in an appendix, together with a historical summary of Theories of Heredity. There are many, of course, who will regard parts of the chapters dealing with Mendelism as consisting largely of speculation; I can only reply that I regard the facts referred to as established, and the theoretical deductions from them as the only ones that have yet been offered which can fit them adequately.

No attempt has been made to quote authorities for every statement, but a list of books and papers is given in which a further account will be found of the subjects treated. The numbers in square brackets [ ] in the text refer to this list. I have also followed the example set by Mr Lock in providing a glossary of unfamiliar terms of which the use has been unavoidable.

For the chapter on Statistical Study of Inheritance my chief sources of information have been Prof. Pearson's *Grammar of Science* and his numerous papers on the subject published by the Royal Society and in *Biometrika*. I have tried to summarise in words the results of his work, and of that of other workers on similar lines, and if the inadequacy of my mathematical knowledge has led me into any serious errors in the attempt, I owe them my apologies.

Chiefly of course I am indebted to Prof. Bateson's recent work on *Mendel's Principles of Heredity* as the most complete and authoritative account of the subject, from the acknowledged leader of the Mendelian school. For permission to reproduce several figures from this book I tender my thanks. In addition to information from many original papers, I have also not hesitated to make use of Lock's *Recent Progress in the Study of Variation, Heredity and Evolution*, Thomson's *Heredity*, and some other books dealing with the general aspects of the subject; to these authors, and to several friends who have been kind enough to give me written information on matters with which they are especially conversant, I wish to record my indebtedness. I wish also especially to thank Mrs A. C. Seward for drawing the sections of *Primula* reproduced in fig. 10 (p. 81).

L. DONCASTER.

CAMBRIDGE,

June 1910.

PREFACE TO THIRD EDITION

IN correcting this little book for a third edition I have made as few changes as possible. Apart from a few slight alterations in the earlier chapters, these consist almost entirely of additions, made necessary by the progress of investigation, to Chapters VII, VIII and IX, and to Appendix II. Without adding materially to the size of the book, it is impossible to deal adequately with the great contributions made by Prof. Morgan and his school to our knowledge of the mechanism of Hereditary Transmission, but I have attempted in Appendix II[1] to indicate something of his methods and conclusions.

L. DONCASTER.

LIVERPOOL,
  March 1920.

[1] In consequence of the death of the author, it has not been possible to make the contemplated additions to Appendix II. For an account of these advances, the reader must be referred to the works of Morgan given in the literature list at the end of this volume.

CONTENTS

CHAP.

I. **Introduction.** Relation of Heredity to other branches of knowledge.—The questions to be answered **. page** 1

IT. **Variation.** Occurrence and kinds of Variation.—Continuous variation and methods of study.—Discontinuous variation.—Inborn and acquired characters . 7

III. **The Causes of Variation.** Mutation.—Action of environment on body and germ-cells.—Variation on crossing.—Relative importance of 'inherent' and 'acquired' characters 22

IV. **The Statistical Study of Heredity.** Two methods of studying Heredity. —The biometrical method.—Correlation and Regression.—Parental correlation and the Law of Ancestral Heredity.—Heredity in 'pure lines' (Johannsen etc.).—Inheritance of 'mental and moral' characters in Man.—Eugenics 32

V. **Mendelian Heredity.** Mendel's Law illustrated.—Segregation and Allelomorphism.—Crosses concerning more than one pair of characters.—Examples of Mendelian characters in Plants and Animals.—Combs of Fowls.— The Andalusian Fowl 52

VI. **Mendelian Heredity** (*continued*). **The Inheritance of Colour.** Concurrence of two factors in the pro-

CHAP.

duction of Colour.—Colour in Animals; some colours 'epistatic' over others.—Flower-colours; Reversion on crossing.—The nature of albinism.—More complex cases of interrelation; Stocks, Primulas   .   .   .   71

VII. **Some Disputed Questions.** Mendelian segregation.— Inheritance of acquired characters.—Indirect and experimental evidence. — Telegony. — Maternal Impression   85

VIII. **Heredity in Man.** Physical and Mental Characters.— Diseases.—Mendelian Characters; Eye-colour, Brachydactyly, abnormalities of the Eye. — Non-Mendelian characters; Skin and Hair-colour. — Importance of Heredity in relation to Sociology   .   .   .   104

IX. **Heredity and Sex.** Secondary sexual characters.— Dominance altered by Sex.—Sex-limited inheritance; the Currant Moth, Fowls, the fly *Drosophila*.—Sex-limited affections in Man; Colour-blindness, Night-blindness and Haemophilia.—Sex inherited as a Mendelian character.— Sexual Dimorphism   122

APPENDIX I. **Historical Summary of Theories of Heredity.** Lamarck.—Darwin and the Theory of Pangenesis.— Weismann's Theory of Germ-Plasm   .   .   .   138

APPENDIX II. **The Material Basis of Inheritance.** The Nucleus and Chromosomes as possible 'bearers' of Heredity. Behaviour of Chromosomes in Germ-cell formation   148

LITERATURE LIST   154

GLOSSARY   158

INDEX   162

CHAPTER I

DURING the whole history of scientific enquiry, one of the most fascinating and at the same time one of the most baffling of the problems which confront mankind has been the cause of the resemblances and differences between parents and children. In general, the facts are common knowledge; the essence of Heredity and Variation is expressed in the proverbs 'Like begets like' and 'Nature never uses the same mould twice.' Yet clearly the two proverbs are contradictory, for if like really begets like Nature must use the same mould for all the members of a family. Our object therefore is to investigate, first, how the characters of a parent actually are distributed among the children, and how the offspring of the same parentage may differ among themselves; and secondly, if possible, what is the mechanism by which the resemblances and diversities are brought about.

These problems are interesting from various points of view. They attract us for their own sake, as does anything mysterious or unexplained; they have a deep human and practical importance, for not only do they affect us all individually, but upon their solution depends, to an extent as yet only dimly realised, the answer to some of our most pressing social questions; and finally they lie at the very root of all theories of organic evolution, so that they form as it were the basis of philosophical biology. The relation of the study of Heredity and Variation to sociology must be left to a later chapter, but before proceeding further we must shortly consider its bearing on theories of evolution.

The fact of organic evolution is admitted by all schools of biology, but about the causes of the process and the manner in which it takes place there is still wide diversity of opinion. To some of the more important theories of evolution it will be necessary to refer again later, but however great may be the difference of opinion with regard to them, all biologists agree that evolution depends ultimately on Variation and Heredity. Darwin called his great book *The Origin of Species* because the unit step, so to speak, on the scale of evolution is the transition from one species to another. But if a species A is to give rise to a species B, in the first place some individuals of A must vary in the direction of B, and then the variation must be inherited, for otherwise no permanent change can take place. The differences with regard to the cause and method of evolution arise therefore partly from our ignorance of the laws of variation and heredity, and partly from different ideas as to the causes which lead to progression in certain directions rather than in others. This latter source of disagreement is to a large extent outside the province of this book, but the subjects of Heredity and Variation are so intimately bound together that one cannot be adequately treated without the other. If, however, we can come to any definite decision with regard to the nature of Heredity and Variation, we shall have made a long step towards understanding the method by which evolution has taken and is taking place.

One other point must be mentioned. The study of heredity brings us face to face with perhaps the most fundamental problem of biology—the ultimate nature of living matter. For if an ovum, barely visible to the eye, or the much smaller spermatozoon which is visible only with high magnification, can bear potentially all the parental characters which may be inherited by the offspring, it is clear that any hypothesis of the nature of living matter must take these things into account; and though we cannot unravel or even imagine it, we can at least get some idea of the amazing complexity of the substances which in thoughtless moments we group together under the single name of 'protoplasm.'

We will now attempt, by means of a few examples, to illustrate some of the questions which must be answered, and some of the facts which must be brought into relation, by any consistent account of the process of heredity. A tall man on the average has taller children than a short man, but if all the sons of a number of tall men were measured, it would be found that they showed every gradation in height between the tallest and shortest ; some would be taller than the fathers, others shorter, but every gradation between them would occur. Also, if a tall man marries a short wife, the sons are neither all as tall as the father, nor divided sharply into a tall group and a short group ; again they make a graded series from short to tall. But if we cross a tall variety of the sweet-pea with a dwarf variety, all the offspring are as tall as the tall parent, and among the offspring of these crossed talls, some are tall and some short, but none are intermediate. Here then we get two distinct modes of inheritance, and also two kinds of variation ; in the first case the character varies in such a way that all intermediates are found between the extreme conditions, and in the second the individuals can be classified sharply into two groups. Again, we cross a white mouse or rabbit with a black one, and all the offspring may have the grey-brown colour of the wild animal—we have produced what is called reversion to the wild type, and have obtained a form different from either parent. But if we mate the same black parent with another white individual, it may happen that all the offspring are black, and instead of reverting to the wild form they all follow one parent. If either the greys or blacks produced in this way are mated together, some of their young will be white; although none of the children of the original white individual resembled their white parent in colour, yet the white has appeared again among the grandchildren after skipping a generation. In man, a colour-blind father rarely has colour-blind children, but some of his nephews and male grandchildren through the female line are usually affected; that is to say, the disease appears in males but is transmitted by females.

It is clear from this short list of examples that there are a number of different forms of hereditary transmission, and our object must be, first to classify them into groups in which the behaviour is similar, and next to attempt to bring them under a common scheme. And it is also clear that the different kinds of heredity are associated with different kinds of variation; for example variation in height in man is inherited differently from variation in colour-vision, and both differ from variation of coat-colour in rabbits, in their inheritance.

A question of a different kind is the cause of inherited differences, and whether differences due to

the action of circumstances are inherited. Does a
man, for instance, who develops certain muscles by
frequent use, or who injures his health by excessive
drinking, have children with larger muscles or poorer
health in consequence? The question is frequently
answered in the affirmative, but it is part of the
province of the study of heredity to investigate the
matter, and in these and all other cases to decide
not only whether a character is inherited, but, if it is,
to what extent and in what manner it will appear in
the offspring.

CHAPTER II

VARIATION

WE have seen that the subjects of Heredity and
Variation are so closely connected that one cannot
be considered apart from the other, for without
variation all the offspring of the same parents would
be exactly alike, and the study of heredity would
resolve itself into an investigation of the cause of
this likeness. But the actual problem is much less
simple; it includes the questions how and why the
members of a family may differ from one another,
and according to what rules and by what means
these differences are transmitted to later generations.
In practice therefore the study of heredity is the
study of the manner and cause of the inheritance of
variations, and hence the nature of variation must
be examined before enquiry into its transmission.

Before the time of Darwin variations were frequently regarded as abnormalities, inconvenient to
the systematist and of relatively small importance.
Every species was supposed to conform to the type

originally created, and divergences from this type
were regarded as imperfections. But it was obvious
that there was always more or less fluctuation about
the type in different individuals, and breeders of
plants and animals made use of this want of uni-
formity to select the best specimens and so to
improve the race. The Natural Selection theory
of Darwin and Wallace supposes that a process
comparable with this takes place in nature, and so
brings about the adaptations of natural species.

Of the causes which induce variation nothing
definite was known, but Darwin's belief was generally
accepted that it is due to changes in environment
acting directly or indirectly on the organism. He
regarded the action of such changes as cumulative
through a number of generations, so that its effect
in producing variation might not be visible until the
change had acted on several generations. This belief
was founded on the observation that animals bred
in captivity appear to be much more variable than in
the wild condition, and the changed conditions of
life are supposed to induce the variation. But species
in nature are not by any means subject to uniform
environment, and thus their variability was ascribed
to similar causes.

Darwin and Wallace pointed out that variation
occurs in all parts of every species, that it appears
to occur in every possible direction, and to every

extent from very small to considerable range. They therefore founded their theory on this type of variability rather than on the occurrence of considerable 'occasional variations' which are not connected with the type by a series of intermediates. It was not, however, until after the theory of Natural Selection had obtained general recognition, that any detailed study was undertaken of the actual frequency and extent of variation, and its mode of occurrence.

The accurate investigation of variation has thus been in progress only for some twenty or twenty-five years, and according to the methods adopted students have become divided into two somewhat distinct schools. One of these has devoted itself rather to the attempt to observe and classify the different kinds of variation, and the other, generally called the 'biometrician' school, to measure its frequency and range. It will be convenient to consider the results obtained by the second method first.

If a character is chosen which can be accurately measured, such as human stature, and a sufficiently large number of individuals are observed, it will commonly be found that there is considerable range of variation, and that every gradation in size occurs between the smallest and largest. Such variation is spoken of as 'continuous,' as opposed to 'discontinuous' variation in which individuals of two kinds occur, which are not connected by intermediates. Further, in cases of continuous variation it will appear that one size is more common than any other, and, in the simplest cases, that the individuals are progressively rarer as the size of the structure considered diverges more and more from the most frequent value. The most frequent condition is named the 'mode,' and its

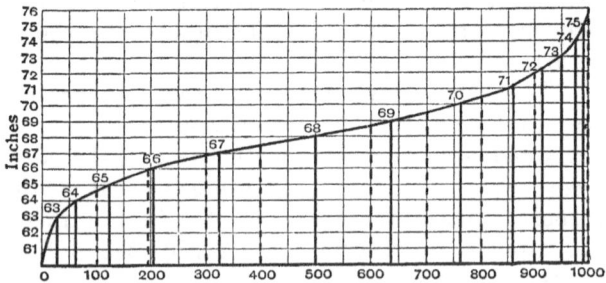

Fig. 1. Curve illustrating stature, the vertical scale representing heights above 60 inches, the horizontal scale numbers of individuals who are supposed to be placed side by side in order of their height.

size the 'modal value' for the character. For example, if the heights of a large number of men were measured, it might be found that they ranged by every gradation from 60 to 76 inches. If the measurements were taken to the nearest inch, it might then be found that a greater number had a stature of 68 inches than any other height, that the next most

frequent heights were 67 and 69 inches, and that the
more the stature differed from 68 inches in either
direction, the fewer would be the men having that
measurement. This could be represented graphically
by arranging vertical lines representing the heights
of every man in order of their height; a line joining
their tops would then rise rapidly at the lower end,
would be nearly flat as it passed over the men having
heights near the 'modal value' of 68 inches, and
would rise again steeply to the exceptionally tall
men at the upper end of the row (fig. 1). [13][1]

A more instructive method of graphically representing the distribution of variation is to take a base-
line and divide it into equal parts, each representing
an equal increment in the structure measured. From
each division of the base-line a vertical line is drawn
representing by its length the number of individuals
having that measurement.

In the imaginary case taken above, the base-line
would have 17 divisions representing successive
heights of from 60 to 76 inches; at each division
a vertical line is drawn which by its length represents the percentage of the population which have
that height (fig. 2). By joining the tops of the per-
pendiculars (ordinates) a curve, or more strictly a
polygon, is obtained which graphically represents the

---

[1] For references see the end of the Volume.

distribution of the variation among the population
measured.   The highest point of the curve represents
the mode for the character, and the extremes of
variation are where it touches the base-line.   The
more numerous are the subdivisions into which the

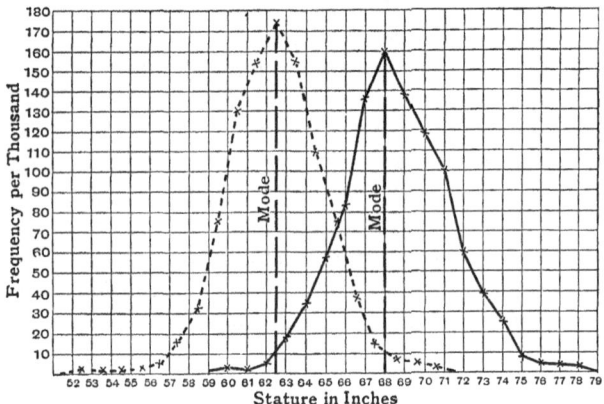

Fig. 2.   Curves showing distribution of stature in women (mothers)
—dotted line; and in men (fathers)—continuous line. The curves
approach the 'normal curve.'   (Data from Pearson.)

variable character is classified, the more nearly the
line joining the tops of the ordinates will approxi-
mate to a smooth curve ; e.g. if the population were
measured to the nearest quarter of an inch instead
of to the nearest inch there would be four times

as many ordinates and the curve would be nearly smooth.

It is clear that a curve of this kind can be used for comparing the variability of different characters, for the greater the variability of the population the wider will be the base ; consequently the curve for a very variable character will be relatively low and wide, that for a slightly variable one measured in the same scale will be tall and steep. A curve of this kind, which is quite similar on either side of the longest perpendicular ('median,' representing the modal value), may be obtained by plotting any measurements which vary fortuitously around a most frequent value, and such a curve is called a 'normal curve.' For example, if a large number of beans in-cluding equal numbers of white ones and black ones were placed in a sack, and drawn out ten at a time without selection of colour, most frequently five white and five black would be drawn, less often six of one colour and four of the other, more rarely seven and three and so on to the rarest case of ten of one colour. If the numbers of white beans in a draw are plotted along the base-line, and the ordinates represent the number of draws for each combination, a polygon approaching the normal curve will be obtained. Variation which gives a normal curve when plotted in this way is spoken of as normal variation.

As mentioned above, the steepness of the curve is

a measure of variability, and this can be expressed by taking a point in the curve, the perpendicular from which to the base-line divides the area, enclosed by the curve, the median and the base-line, into two equal parts. Or, differently expressed, the perpendicular divides the curve in such a way that the number of individuals between it and the mean is the same as that between it and the extreme. The distance of this perpendicular may be used as a measure of the variability of the character considered, for clearly the greater the variability (and thus the flatter the curve), the further this perpendicular will be from the median[1].

In many variable characters, the frequency of variation below the mode is not exactly equal to that above it, in which case the curve will be steeper on one side of the mode than on the other, and the average value for the character ('mean') will not be identical with the mode. For example, if the variation in the number of children in a family were plotted in this way, the sizes of families would range from 0 to about 20, but the most frequent number would perhaps be four. Four would then be the

---

[1] In practice, not this perpendicular, but another rather further from the median is used, which for practical purposes is more convenient. The distance of this perpendicular, measured in units of the horizontal scale, is called the 'standard deviation' and is regularly employed as a measure of variability.

modal value, but the average or mean might be about six ; the curve would rise steeply to the mode, and fall away more gradually to the maximum number (fig. 3). Such a curve is described as 'skew.' In

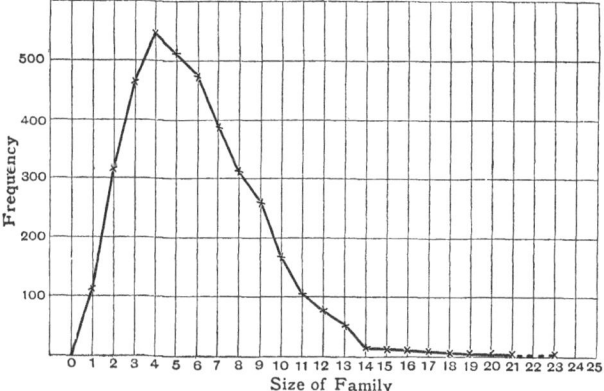

Fig. 3. Curve showing distribution of size of 3837 families in America containing deaf-mutes. (After Schuster, 'Hereditary Deafness.' *Biometrika*, Vol. IV. 1906, p. 474.)

extreme cases the mode is at one end of the curve, when variation takes place only on one side of it, e.g. in the marsh-marigold, the most frequent number of 'petals' is five, but there may rarely be six, seven or eight, but practically never less than five, so that in plotting the frequency a 'half-curve' is obtained (fig. 4).

Another rather frequent condition is that the curve has two maxima or modes (fig. 5), indicating that a large number of individuals have a low measurement, a less number are intermediate, and again a larger number have a higher measurement. Species which vary in this way are called 'dimorphic,' or if

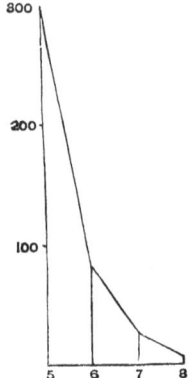

Fig. 4. 'Half-curve' representing the number of 'petals' on 416 flowers of the marsh-marigold (*Caltha palustris*). (After de Vries.)

there are more than two peaks to the curve, 'polymorphic.' Further, it is possible that the two parts of the curve should be entirely separate, if intermediates between the low and high groups are completely wanting. This type of variation is spoken of as 'discontinuous' in contrast to the 'continuous' variation hitherto considered. It is possible that dimorphic cases in which intermediates exist are really essentially discontinuous, but that the two groups into which the species is divided each exhibit continuous variation about the mode for the group, to such an extent that the higher members of one

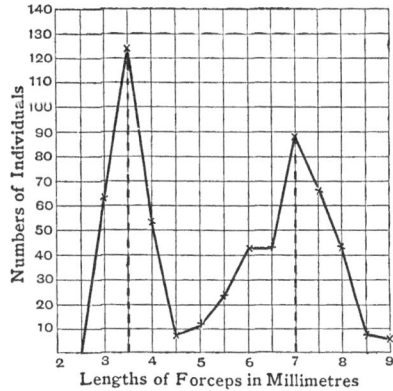

Fig. 5. Curve with two modes, representing frequency of lengths of forceps of male Earwigs from the Farne Islands. (After Bateson.)

group overlap the lower of the other. For example, if the modal (most frequent) stature for a race of men were 68 inches, and for the women 62 inches, it might happen that on plotting a frequency curve for the stature of adults including both sexes, a curve would be obtained having two maxima, one near 62 and another near 68. Yet the stature might be a definite sexual character, and hence essentially as discontinuous as the sexes themselves. This distinction between continuous and discontinuous variation may seem unimportant in itself, but when its inheritance is considered the distinction becomes of the first importance.

Clearer examples of discontinuous variation are given by such characters as colour, or by organs which are repeated in series, such as vertebrae and ribs, the segments of a worm or the petals of a flower. When variation occurs in this latter group it is generally complete, so that the different forms are visibly discontinuous. In the case of colour in the skin or hair in animals, or petals of flowers, discontinuity is sometimes less apparent, and grading frequently occurs, but even in apparently graded cases the inheritance of the character may often reveal discontinuity. For example, a piebald animal might be thought to be intermediate between the fully-coloured and albino (white with no pigment), but breeding tests would at once show that piebaldness was an independent character, which cannot be regarded as in any sense intermediate between the other two conditions except in general appearance. The same applies to such cases as the 'silver' cat or rabbit, or the pale purple sweet-pea ; the cause of the pale colour is entirely distinct from the cause of the absence of pigment in the white varieties of those species.

The recognition of the importance of discontinuity in variation, which we owe chiefly to the work of Bateson in England and De Vries in Holland, is one of the chief advances which the study of the subject has made since the time of Darwin.

One other distinction between different kinds of variation must be mentioned here, which will be discussed more fully in subsequent chapters. The kinds of variation mentioned above are all *inborn*, or inherent in the individual and to a great extent independent of its manner of life. But it is well known that the continued use of an organ or structure, or the prolonged action upon it of some external stimulus, may alter its form or cause it to assume a condition different from that which it would have had if these influences had not acted. In general, an organ tends to adapt itself either to the uses to which it is put or to the action of the environment which surrounds it. The muscles of a limb used for strenuous work increase in size and strength, or a part of the skin continually exposed to bright light develops a deeper colour than if it is covered. The converse process is also true ; an organ which is not used or exposed to its normal stimuli tends to diminish, and become less

2—2

adapted to the use to which it is normally put. Such characters as these, arising in response to a stimulus, and not appearing in its absence, are technically called 'acquired characters,' a phrase which it will be necessary to use rather frequently in the following pages. As a rule, such 'acquired characters' are adaptive, that is, they render the organism or structure better fitted to its surroundings than if they had not been developed. The older students of heredity never doubted that these acquired characters were inherited as strongly as the inborn characters discussed above, but since the publication of Weismann's theory of heredity (see Appendix I) with the great body of evidence which he has collected on the other side, opinion has turned increasingly towards the belief that acquired variations are not transmitted. Weismann regards the germ-cells as essentially distinct from the rest of the body, so that acquired modifications of the body cannot be transmitted because the germ-cells are not affected. The germ-cells collectively, or rather that part of them which is concerned with the transmission of hereditary characters, he calls 'germ-plasm,' the rest of the body consisting of 'body-plasm' (or 'soma'), and he regards 'acquired' modifications as affecting body-plasm only. A developing germ-cell gives origin to both germ-plasm and body-plasm of a new individual, and hence characters borne by the germ-plasm appear in the body; but since body-plasm cannot be converted into germ-plasm, modifications of the body cannot be transmitted to offspring.

The possible inheritance of acquired characters is treated more fully in a later chapter.

CHAPTER III

In the last chapter the distinction has been explained between continuous and discontinuous variation ; some confusion has however arisen with regard to the terms used in describing these conditions. Continuous variation about a mean (or more accurately modal) condition is sometimes spoken of as 'fluctuation,' but as will be seen below this kind of variability probably includes two very distinct groups of facts. It may include inherent variability arising in the germ-cells, or it may include differences in the adult condition having their origin in different effects of environment during growth. Some writers have used the word 'fluctuation' for this latter condition only.

Discontinuous variation is sometimes called 'mutation,' a word which also has been used in several senses. It may mean the appearance of a form varying discontinuously from the type, or it may be applied to the discontinuous character itself.

A more serious source of confusion is that the term is used by some to denote any discontinuous variation arising 'spontaneously,' by others for cases in which the variety differs from the type in several apparently distinct characters, and not only in one, so that the new form constitutes an 'elementary species.' Since in studying heredity it is usually important to consider distinct characters separately, it may be permissible to use the word for the origin of a form differing recognisably from the type and not connected with it by true intermediates.

It has already been pointed out that very little is accurately known about the causes of variation, and it is not impossible that the different forms of variation have different origins. Most writers agree that the ultimate cause must lie in the action of environment in some form, but as Darwin clearly stated in the *Origin of Species* the environment may act directly or indirectly. In variation of size for example, it is clear that the supply of nourishment, etc., during growth may have considerable influence on the size of the adult, and such variation will commonly be continuous owing to the evenly graded action on different individuals. In these cases the action is direct. If, however, Weismann's theory of germ-plasm and body-plasm is correct, such action may affect only the body and not be transmitted to offspring. It is also possible that the germ-cells may be indirectly affected, giving rise to variation in the offspring; in such a case, however, there is no necessity that the effect on the offspring should be in any way similar to the direct effect of the conditions on the parent. Nothing is known of the nature of possible effects of environment on the germ-cells; the action may possibly be effective immediately and give rise to variability in the next generation, or it may be that the effects are cumulative and only cause visible changes after several generations have been exposed to the same influences. Galton [13] suggested that the organism may have a certain 'stability,' but that influences acting for several generations may have a cumulative effect which will gradually alter the equilibrium until it is finally upset and falls into a new condition of stability, giving rise to an apparently sudden variation. A chemical analogy may make this clearer. If litmus is added to an alkaline solution its colour will be blue. Acid may now be added drop by drop to neutralize the alkali, and suddenly, when the solution becomes acid, the litmus turns red. Examples of variation of which this may possibly be an analogy will be given below.

With regard to the action of environment on the body many facts are known, but it is not certain that they really have any bearing on the question of the origin of variation. For variations so produced are 'acquired characters,' and in many cases at least there is no evidence that they are inherited. For example, many butterflies have two generations in the year, one of which lives through its whole life-history in the summer and the other passes the winter as a pupa (chrysalis). In some cases the two generations are strikingly different, and it has been shown that by freezing the pupae of the summer brood at the right stage, specimens like the spring brood can be obtained. The difference between the two generations is thus due to the action of cold on the pupa. But the two forms regularly alternate in nature and the effects of cold are not inherited. In plants, some species produce quite different leaves according to whether they are grown in water or in dry soil, but the conditions act on the individual, and do not affect its progeny. In such a case, what is inherited is the faculty of making a certain definite response to definite conditions, and this faculty is present whether the conditions operate or not. In man such diseases as tuberculosis are commonly called hereditary; this however does not mean that the child has the disease because his parent had it, but that the parent had a constitution liable to that disease, and the child inherits a similar constitutional liability. If the parent had never been exposed to infection the child would still inherit the liability, for what is transmitted is not the disease or its effects, but the faculty of acquiring it if exposed. It will be found

that most cases which at first sight seem to support
the theory of the inheritance of acquired characters
are equally explicable in the view that both parent
and offspring are susceptible to the action of the external factor ; what is inherited is not the character
acquired, but the innate power of acquiring it.

But it is always possible that some forms of
external conditions may act on both the body-cells
and germ-cells concurrently, and produce similar
effects in each. For example, it may happen that
extremes of temperature produce striking colour-
variations in certain butterflies. Weismann has
pointed out that, according to his theory, in a
developing butterfly the determinants for producing
colour not only exist in the germ-cells which will
transmit the character to the offspring, but also in
the embryonic cells of the body which go to produce
the coloured parts of the perfect insect. If extremes
of heat or cold cause changes in the colour-deter-
minants in the developing wings, so that abnormal
colours result, it is possible that the determinants
in the germ-cells which transmit the colour-pattern
to the next generation will be similarly modified, so
that the offspring will show similar abnormalities.
This would not be the transmission of an 'acquired
character' in the strict sense of the expression, but
the simultaneous modification of body and germ-cells
in the same manner.

But as mentioned above, it is possible that the same factor acting on body and germ-cells may produce different results in the two cases, so that the individual on which the influences have acted may show one modification and its offspring another. It is also possible that not all the germ-cells will be affected alike, and so among the progeny some will show modification and others not, or some may be differently affected from others; for the conditions of stability of different germ-cells may conceivably be different. Certain experiments on insects give reason for supposing that this is so. The results obtained by Fischer, Standfuss and others from exposing pupae of butterflies and moths to abnormal temperatures, while not entirely concordant among themselves, on the whole indicate that moderate degrees of heat and cold tend to alter in the same way the whole batch of insects treated, often in the direction of varieties of the species naturally occurring in warmer or colder climates. But excessive heat or cold causes extreme variations among only a small proportion of the insects treated, and among the offspring of these abnormal specimens only a small fraction are abnormal, and some of these have not the same abnormality as the parents. These observations, together with the fact that the variations produced by heat, cold, and other disturbing factors, may all be similar, suggest that extreme conditions may upset the stability of the type, causing abnormalities to appear, and that some of the germ-cells may also be altered, but not necessarily in the same manner as the body-cells.

An American zoologist, Tower, describes the production of mutations by the action of environment in a beetle (*Leptinotarsa*). In nature he found about one such variation among 6000 specimens; when bred in captivity they were more frequent, but when the full-grown beetles were exposed to extremes of heat, humidity, etc., during the maturation of the eggs, the offspring may include a large proportion (over 80 per cent.) of 'mutations.' These were of several distinct kinds, like those rarely found in nature, and when bred together they are stated to breed true. In this case the abnormal conditions produced no effect on the individuals exposed to them, for they already had their final form, but as their eggs were matured under these conditions the action took effect on the eggs, and mutation resulted among the offspring. When part of the eggs of an individual were matured under abnormal, another part under normal conditions, mutation occurred only among the offspring in the first case, all the beetles in the second being normal. It should be noted that as in the experiments with butterflies the effect of changed conditions was not specific; the same conditions may produce more than one kind of mutation in the same batch of eggs, and some eggs were not affected at all. In both cases the abnormal environment seems to upset the equilibrium, but the effects may differ in different individuals. It is the nature of the organism or germ-cell affected which determines whether and to what extent the change shall take place ; the environment merely supplies the stimulus.

It will be seen that our knowledge of the causes of variation, in so far as these are connected with environment, is very incomplete and unsatisfactory, for although it is fairly clear that conditions may sometimes disturb the equilibrium of the germ-cells and provide a stimulus to variation, yet we have no knowledge of the way in which the stimulus acts and can make no prediction as to the direction the variation will take. Before leaving the subject, one other cause of variability must be mentioned—the effect of crossing different races in producing variation. It frequently happens that the result of crossing distinct races is that the crossed individuals differ from either parent ; sometimes in the direction of increased vigour, as was pointed out by Darwin, and other more recent observers ; sometimes by the development of characters apparently not possessed by either parent, as in the case of 'reversion on crossing.' The cause of this latter phenomenon will be discussed in a later chapter. In the subsequent generations from the cross great diversity may often appear, and Darwin supposed that the mingling of two distinct germinal stocks had an effect in disturbing the equilibrium similar to that produced by change of environment. To some extent this is doubtless true, but recent developments of the theory of heredity have afforded a more exact explanation, in the recombination of the different characters of the two races which are crossed. A fuller account of 'variation induced by crossing' must therefore be postponed until the principles of heredity have been discussed.

One further question should be mentioned before proceeding to the subject of heredity, namely, the relative importance of 'inherent' and 'acquired' characters in making up the sum of characters of a mature individual. It is often assumed, especially in human cases, that the environment has a preponderating influence in shaping the individual. In a certain sense this is true, for many characters can only develop in a suitable environment; muscles must be exercised to be properly formed and the mind cannot develop its full powers if it is never used. But the study of variation leads inevitably to the conclusion that the inherent characteristics are all-important, and that the effect of environment is not much more than to give them opportunity to develop. This is perhaps most impressively seen in the case of 'identical twins,' as has been shown by Galton [12]. There is reason to believe that such twins are produced by the division of one ovum, and even if exposed to different conditions they remain through life much more alike than ordinary brothers who may be brought up under precisely similar surroundings. The same fact is still further emphasised by the study of heredity.

CHAPTER IV

THE STATISTICAL STUDY OF HEREDITY

In studying heredity, either of two methods may be adopted. We may either choose a character and observe or measure its development in a large number of parents and in their children, and so deduce the average extent of resemblance between parents and children for that character; or we may consider a number of individual cases separately, and endeavour to discover the manner in which the character appears in the children who have parents or ancestors possessing it. With regard to the first method Prof. Pearson has written 'We must proceed from inheritance in the mass to inheritance in narrower and narrower classes, rather than attempt to build up general rules on the observation of individual instances.' And '...the very nature of the distribution ...seems to indicate that we are dealing with that sphere of indefinitely numerous small causes, which in so many other instances has shown itself only amenable to the calculus of chance, and not to the analysis of the individual instance' [25, 'Math. Contrib. III.' *Phil. Trans. Roy. Soc.* A, 1896, p. 255]. The second method on the other hand has been used in cases where the causes of variation appear to be few and definite, and seeks to isolate these causes. The first method is thus clearly adapted especially to characters which vary continuously and which can be measured; the second to characters which vary discontinuously and can be sharply separated into classes. The first method gives on the whole the average intensity of inheritance, but little information with regard to its probable development in individual cases; the second attempts to answer the question in what manner the character will be distributed among the offspring in any family.

The founder of the modern statistical, or as it is now often called, the biometrical study of heredity was Sir Francis Galton, and its leading exponents have been Professor Karl Pearson and the late Professor Weldon. In this chapter an attempt will be made to explain the fundamental principles on which the biometric methods rest, and to outline the chief results obtained; the methods themselves frequently require mathematics of an advanced order, and for the study of them the reader is referred to the books and papers dealing with the subject mentioned in the bibliography.

It has already been seen that in the case of a

character which varies continuously about a mean or
mode, the greater the divergence from the mode in
either direction, the fewer will be the individuals
showing that divergence. In the case of human
stature, if the modal height of a population is 68
inches, there will be fewer men of 64 or 72 inches
than of 66 or 70, and still fewer of 63 or 73 inches.
If now the sons of all the men having a given divergence were measured, and it were found that they
averaged as great a divergence from the mode as
their fathers, it is clear that on the average the
height of the sons would equal that of their fathers.
This does not mean that every son would exactly
resemble his father in stature, but the sons would
vary about the paternal stature equally above and
below it, and when plotted in a curve their statures
would make a curve having the paternal stature as
its mode. The average stature of the sons would
then be completely determined by the stature of the
fathers. If on the other hand the stature of the
father had no relation with that of his sons, it is
clear that the statures of the sons of fathers of any
height would vary about the mean of the general
population considered. In practice it is found that
the modal value for sons of fathers of a given height
is between the height of their fathers and the mode
of the general population. That is to say, if the fathers
diverge a given amount from the general mode, their

sons will on the average diverge less ; they will vary
about a modal value lying between the general mode
and the fathers' measurement. This fact is called
'regression.' It sometimes seems paradoxical to those
who have not considered it that the mean deviation
of children from the general mode is always less than
that of their parents.   But of course it does not mean
that all sons of tall fathers will be shorter than their
fathers ; some will be as tall or taller, but the sons of
a number of fathers of given stature will vary about
a mode lying between the fathers' stature and the
mode of the whole population.

Now it is plain that the amount of regression is a
measure of the intensity of inheritance ; if the modes
for sons of fathers of every deviation have deviations
nearly as great as those of the fathers, the intensity of
inheritance would be high ; if the modes for the sons
deviate but slightly from the general mode, whatever
be the deviation of the fathers, the intensity would
be low.   A definite case may make this clearer.
Suppose the modal stature of the population is 68
inches ; it might then be found that for fathers of 64
inches (i.e. deviating 4 inches below), the height of
the sons ranged from 61 to 72 inches.   If, however,
their modal value had a deviation only slightly less
than the fathers' deviation, say with a mode at 65
inches, the regression would be slight and the intensity
of inheritance high ; if the sons' mode had a deviation

much less than the fathers', say at 67 inches, the regression on the general mean would be considerable and the intensity of inheritance low. If then we can find means of determining the ratio between the deviation of sons in general and the deviation of their parents, we shall have a measure of the intensity of inheritance for the character considered. This ratio is called the 'coefficient of correlation' between father and son for that character. It should be noticed that correlation simply means that two quantities vary in relation to each other; the correlation between parents and children is a convenient method of estimating the intensity of inheritance, but correlation exists between any two related variables, e.g. between the measurements of two limbs in the same individual, such as an arm and a leg, or between the numbers obtained in successive throws of dice, if not all the dice are picked up for the second throw. The correlation between the same measurement in brothers may be used as a measure of inheritance, for two brothers resemble each other more than two chance individuals because they are children of the same parents.

The principle of obtaining a coefficient of correlation between father and sons is as follows. It will be convenient to assume that the variability of the character considered is normal, i.e. that the frequency curve falls evenly on either side of the mode, so that the mode is identical with the mean. Stature in inches may be taken as an example. A large number of fathers and a son of each are measured to the nearest inch; it can then be found what is the average measurement of sons for fathers of each

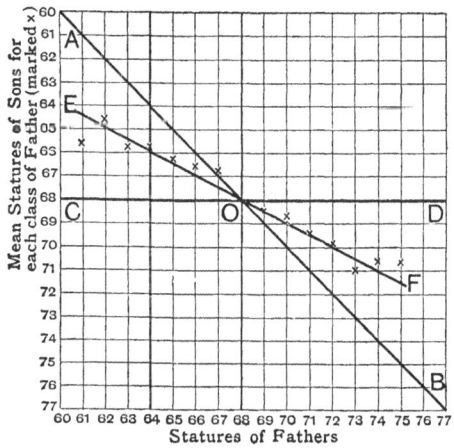

Fig. 6. Diagram of correlation between fathers and sons.

height from the lowest value to the highest. It will be found that the mean deviation of the sons from the mean of the population is less than the deviation of the fathers for each class of fathers. The average ratio between the mean deviations of the sons to the deviations of the fathers is then the coefficient of correlation between father and son for this character[1]. This is more clearly seen in diagram form.

If a square is made with its sides divided into equal lengths corresponding to equal increments in stature from 60 to 76 inches, the top may represent the scale of statures of fathers and the side the scale of mean statures of sons for each class of fathers. If, then, there were complete correlation between fathers and sons, the mean stature of sons of fathers 62 inches high would be 62, of fathers of 63 inches, 63, of 64, 64 and so on. If on the other hand there were no correlation, the means of the sons of every class of father would be the mean of the population (68).

In the first case the line joining the points representing the means of the sons would be a diagonal running from corner to corner (AB), in the second case a horizontal line running across the middle (CD). But if the correlation is between these extremes the line would lie between the diagonal and the horizontal (EF), and the greater the correlation the steeper would be the slope of EF. The steepness of this line is thus a measure of correlation, and since all these lines pass through O in the middle of the square, the

[1] It is assumed throughout that the variability of the sons is similar to that of the fathers. If their variability were different this would have to be allowed for. The variation is also assumed to be normal, so that the mode in each case coincides with the mean.

slope is measured by the size of the angle EOC. The angle made by the diagonal at O is 45°, the tangent of which is 1 (unity). If there were no correlation the angle would vanish, EF coinciding with CD, and the correlation coefficient would be 0. Intermediates are represented by the value of the tangent of the angle EOC. In practice it rarely happens that the points representing the means of the sons for each class of fathers lie in a perfectly straight line ; when they approach it closely the correlation is called 'linear'; when they depart from it considerably, it is called 'skew.'

Table of intensity of parental inheritance in different species. (From Pearson.)

Species	Character	Mean value	Number of Pairs used
Man	Stature	·506	4886
	Span	·459	4873
	Forearm	·418	4866
	Eye Colour	·495	4000
Horse	Coat Colour	·522	4350
Basset Hound	Coat Colour	·524	823
Greyhound	Coat Colour	·507	9279
Aphis	Ratio of right Antenna	·439	368
(*Hyalopterus*	to Frontal Breadth		
trirhodus)	(non-sexual reproduction)		
Water-flea	Ratio of Basal Joint of	466	96
(*Daphnia magna*)	Antenna to Body length		
	(non-sexual reproduction)		

Prof. Pearson and his collaborators have worked out the correlation between parent and child for a number of measurable characters in Man, Animals, and Plants, and they find that the numbers group themselves about a value not far from 0·48, varying from 0·42 to 0·52. That is to say, on the average the offspring deviate from the mean about half as much as the parent.

The parental correlation hitherto discussed has taken no account of the second parent, for if individuals mate at random the one parent may be considered alone, and the second will *on the average* have the mean value for the general population. But it is clear that one may take for the parental value in each class the mean of the two parents (making allowance for any difference in measurement due to sex), and plot the means of the sons (or daughters) against the classes so produced. The value derived from taking the mean of father and mother is called the 'mid-parent,' and the correlation so arrived at would give the measure of resemblance between children and their mid-parents. This is naturally higher than the correlation observed when only one parent is considered; for if both parents deviate in the same direction from the mode of the population, the children will average a greater deviation than if only one does so, and still more than if one deviates in one direction, the other in the opposite. We thus obtain a measure of the amount contributed to the offspring by the two parents together, but even now we do not find the correlation complete (1·0) because the contributions from previous ancestors have also to be taken into account.

Galton was the first to introduce the idea of the 'mid-parent,' and he went on to attempt to estimate the average contribution to the children from each generation of ancestors. Since the correlation between offspring and mid-parent is not complete, part of the heritage, which is not visibly present in the parents, must be contributed from more distant ancestors. Galton concluded from the data he collected that on the average half the heritage of an individual may be taken as derived from the two parents, one quarter from the four grandparents, one eighth from the great grandparents, and so on, the whole series ($\frac{1}{2}$, $\frac{1}{4}$, $\frac{1}{8}$, $\frac{1}{16}$—) adding up to unity. Pearson estimates the average correlation between offspring and one parent, as about ·5, of offspring with a grandparent as ·33, with a great grandparent as ·22, the correlation coefficient with an ancestor of each generation being $\frac{2}{3}$ of that of the next below; these numbers, however, are not in any way comparable with Galton's series ·5, ·25, ·125, etc. Galton attempted to estimate the amount of the heritage received from the 'mid-ancestor' of each generation independently of what was received from other generations; but in the metaphor of bequests of property, he calculated that of the total heritage of an individual, half on the average was bequeathed by the parents, one quarter by the grandparents direct to the grandchild and so on. Pearson's series ·5, ·33, ·22 etc. gives the average measure of resemblance between children and an ancestor of each generation, which is clearly a totally different thing. From this series he has worked out figures corresponding to Galton's, making the series ·6244, ·1988, ·0630, i.e. he finds that the parental bequest is greater and the ancestral bequests less than Galton estimated. From the results obtained first by Galton and later by Pearson has been formulated the 'Law of Ancestral Heredity,' which has been stated in various forms, perhaps the most general being 'the mean character of the offspring can be calculated with the more exactness, the more extensive our knowledge of the corresponding characters of the Ancestry' (Yule [44]). But it should be noted that there is an important difference between Galton's original statement of the law, and the later statements of Prof. Pearson. Galton wrote that 'the two parents between them contribute on the average one-half of each inherited faculty, each of them contributing one-quarter. The four grandparents contribute between them one-quarter, or each of them one-sixteenth ; and so on.' He regarded this as a physiological statement of the way faculties were transmitted, while Pearson, in his later writings at least, regards the law simply as a statistical description of what is found when large numbers are observed in mass.

It has been mentioned that the characters which especially lend themselves to statistical treatment are those which vary continuously and which can be accurately measured, but Prof. Pearson has applied similar methods to discontinuous characters, which can be classified into groups but not measured, for example coat-colour in horses. He finds as the results of his enquiries that the inheritance of such characters can be stated in terms similar to those obtained with measurable characters, so that the principle of ancestral correlation leading up to the law of ancestral heredity may be applied to these characters also. But whatever may be the case with characters which vary continuously, it will be seen below that discontinuous characters are commonly alternative in their inheritance, i.e. there is no blending, but the offspring exhibit one or other only; and in some at least of these cases, the character of the offspring cannot be calculated with any more exactness if the ancestry is known than if it is not. Such instances show clearly that although the law may be statistically true when applied to considerable populations, it gives us no clue to the physiological processes which determine the transmission of characters from one generation to another.

Another argument that has been used against the physiological validity of the law of ancestral heredity is based on the work of Johannsen and others who have obtained results similar to his in other cases. Johannsen worked at the inheritance of weight of seeds in beans and in barley, and self-fertilised the plants investigated for a series of generations so as to isolate what he calls 'pure lines.' He found that in beans, for example, the seed-weights of a mixed population gave a normal frequency curve—the weights varied continuously and evenly about a mean value. The beans on an individual plant when the flowers are self-fertilised also form a normal curve about a mean, but this mean is not necessarily identical with that of the race in general. If now the flowers on such an individual are self-fertilised, and the beans produced are sown, the mean weight of the beans on all the daughter plants will be identical with the mean of the beans on the parent, i.e. among the offspring produced by self-fertilisation there is no regression towards the mean of the race. It thus makes no difference whether large or small seeds are chosen *within the pure line* ; the mean weight of the seeds on plants grown from the smallest and largest of the parental beans (seeds) is in each case equal to the parental mean. Selection, therefore, within the pure line has no effect in altering the mean weight of the seeds, for the differences in seed weight within the line are not inherited. The probable cause of this is that the differences between the seeds on a self-fertilised plant are due to the action of external circumstances; the position of the beans in the pod or the position of the pods on the plant cause differences in the nutrition which allow some beans to grow larger than others. These differences are 'acquired characters,' and we have here additional evidence that such are not inherited. It is to variation of this type that the term 'fluctuation' is applied by some authorities.

It is clear then that if selection is made among beans harvested from a mixed population, on the whole the larger beans will belong to pure lines having a higher mean, and thus selection for a few generations will isolate pure lines having a high value, and the mean of successive generations will rise until the largest pure lines have been isolated. Beyond that point further selection will have no effect. This is precisely the result arrived at by Prof. Pearson from a study of selection within a mixed population; the mean will rise rapidly on the first selection, more slowly later, until in very few generations it reaches a point at which selection has no appreciable effect. Pearson calculates that if selection now ceases, the selected race will very slowly revert towards the mean of the general population. But, as has been seen, this conclusion is based on the assumption that continuous variation is due to the concurrent action of an indefinite number of small independent causes. If, however, Johannsen is correct, we may divide these causes into two classes: the causes which induce 'fluctuation' as explained above, which agree with Pearson's requirements, and the cause or causes which give rise to the difference between one pure line and another. Now this second group may conceivably consist in a single factor of the nature of a small 'mutation,' and if so, by isolating the pure line this factor is also isolated, and no return towards the mean of the general population need take place. According to Johannsen this isolation can be effected in one generation by selecting the self-fertilised plants which have the highest average yield, instead of selecting the heaviest beans themselves.

We thus obtain by experiments such as those of Johannsen a new conception of the possible nature of continuous variation; it may be due partly to 'fluctuation' brought about by the action of environment and not inherited, partly to a series of small stepwise 'mutations,' each of which owing to fluctuation overlaps the next, and can only be isolated when it is possible to breed pure lines. It should be said that there is as yet no certainty that this account of continuous variation is sufficient to cover all cases ; it is a suggestion of possibility rather than a statement of fact.

We have seen that there is reason to believe that the Law of Ancestral Inheritance is true only when applied to a large number of individuals considered in mass, or, as it has been put, that it is a statistical rather than a physiological law. In individual cases it is not true that the offspring need be influenced by ancestors beyond the parents, but in other cases, as will be seen in dealing with Mendelian heredity, these ancestors have important effects, so that statistically it is possible to say what is the average influence of the ancestors of any generation upon the offspring. Now in cases where it is possible to define rigidly single characters, much more is learned from the physiological than from the statistical method, but where no such rigid separation of characters is possible the statistical law is the only one that can be applied. This is particularly the case in characters which vary continuously, or where the categories into which the character falls overlap one another, as for example in Johannsen's beans. Further, the statistical method is frequently the only one which is available when experiment is impossible and when our knowledge of the facts is based solely on numerical data from observed cases, and this of

course applies especially to inheritance in Man,
where experimental evidence is not available.

By collecting family histories of distinguished
men, Galton showed long ago [15] that exceptional
mental qualities were inherited; and this work
has recently been much extended and made more
definite by Professor Pearson and his school. It
is commonly believed that exceptionally gifted men
do not have distinguished sons, but this like many
other popular beliefs is only partly true. It
has been seen that if an individual deviates a
certain amount from the general mean, his children
will on the average deviate less, because when the
whole ancestry is taken into account, the effect of
previous generations is to cause regression on the
mean of the population. And since the theory of
regression depends on the assumption that variation
is due to the existence of a large number of independent causes acting concurrently, it is unlikely
that among the limited number of offspring of one
exceptional man any one child will unite in himself
the same combination of factors as went to make up
the father's character. Further, it is improbable that
an unusually gifted man will marry a wife equally
gifted in the same manner, and the mother's influence
on the children is closely similar to that of the father.
It cannot therefore be expected that all great men
should have equally great sons, but they are far more

likely to have exceptional sons than are mediocre men, and if the mother is also exceptional in the same direction this probability is greatly increased.

In the last few years the intensity of inheritance in such characters has been given numerical expression. Professor Pearson, after working out the statistical laws of inheritance in many physical characters of man, animals and plants, has applied the same methods to what are called the mental and moral attributes. Characters were chosen such as vivacity, popularity, conscientiousness, temper, ability, hand-writing, which were estimated by reports from school-teachers on the children in their schools; and also intellectual ability as shown in university examinations or by the position in a public school at a particular age (Schuster [10])[1]. All these, when investigated by the same methods as were devised for the coat-colour of horses or eye-colour in man, are found to give results closely in accord with those obtained for physical features. The conclusion is therefore reached that not only bodily characters, but also those of the mind are essentially determined by the hereditary endowment received from the parents. This result is of great importance practi-

---

[1] In these characters the resemblance between parent and child cannot of course be estimated directly, but it has been pointed out above that the resemblance between brothers may be used as a test of the intensity of heredity.

D.   4

cally ; it shows how little room is left in the
development of the individual for the effects of
environment even on the intellect or mind in the
broadest sense of the word ; no doubt the direction
which intellectual development takes is to a considerable extent determined by circumstances, but
the kind of mind is irrevocably decided before the
child is born. Still less is there room for the
inheritance of the mental acquirements made by the
individual during his life, and hence the hopes held
out of improving the race by education and by
special care of the dull or feeble-minded are illusory,
except in so far as they improve the *tradition.* Just
as the welfare of the race may be increased by an
invention which is handed on from generation to
generation, so the good effects of education or other
improved conditions may be handed on, but this is
not heredity. The father may educate his children
because he himself was educated, but the mental
powers of his children will be the same whether he
had a good education or none[1]. And the effects of
special care given to the weakly or feeble-minded
may be absolutely harmful to the race, if the im-
provement so effected leads to more frequent

[1] Of course education is a necessary condition for the full development of the mental powers, but at present we have no evidence that
it can add potentialities not present at birth. The subject is more
fully discussed in Chapters VII and VIII.

marriage among such unfortunates than would otherwise be the case, for then an increased number of
defective children may be born, and the race-average
be lowered. Hence has arisen the study known as
'Eugenics,' the study, that is, of the methods by
which the race may be improved both physically and
mentally. The whole trend of the results obtained
is that in order to produce exceptionally gifted men
in both body and mind, those with high development
of the characters desired should be encouraged to
marry; and that to prevent the production of the
weakly and feeble-minded, the only method is to
prevent such from having offspring. It is admitted
that at present these things hardly come within
'practical politics,' but there is little doubt that the
nation which first finds a way to make them practical
will in a very short time be the leader of the world.

CHAPTER V

MENDELIAN HEREDITY

In the last chapter the distinction has more than once been referred to between the statistical rules of inheritance discovered by observing great numbers of cases taken together, and the physiological laws which determine the actual manner of transmission in individual cases. The province of the present chapter is to indicate the methods by which one at least of these physiological laws has been investigated, and the results to which such work has led. In studying this part of the subject it is necessary to consider, at least in the first place, characters which vary and are inherited discontinuously, so that they may be sharply marked into distinct categories. The foundation of the study was laid by Johann Gregor Mendel, a monk of the monastery of Brünn in Bohemia. His most important paper was published in 1866 [2], but perhaps owing to the fact that the biological world was then occupied almost solely with the discussion of the 'Origin of Species,' his work attracted no attention at the time, and only became celebrated on its rediscovery in 1900. One cannot avoid speculating on the possible effects on biological thought, had the experiments and conclusions of his now famous contemporary ever come to the knowledge of Darwin.

The method which led Mendel to his great discovery was to experiment with plants exhibiting discontinuous characters, and to consider each character separately. Previous workers in the same field had made many laborious experiments in crossing different races of plants or animals [7], but had always regarded the individual as the unit, and hence arose the belief that mongrels or hybrids were usually intermediate between the parents, resembling one in some features, the other in others, but with no regular rule ; and further, that when hybrids were bred together the offspring were often almost infinitely variable, extending in a series from some closely approaching one original parent through a diversity of intermediate or new forms to others like the second parent. So grew up the belief that the crossing of distinct races or breeds is a potent cause of variability, which, however, except when 'reversion on crossing' took place, seemed to fall under no ascertainable law.

Mendel's most important experiments were made with races of the edible pea, which he grew in the garden of his monastery. He found in peas several characters which vary and are inherited discontinuously, and he crossed together races which differed in one or more of such characters, but in the offspring and later generations he considered the distribution of each character by itself, quite apart from the other characters of the plant. As an example we may take the character height or tallness. Certain varieties of peas grow stems some six feet in height, others are short and do not exceed about two feet. The heights fluctuate about a mode, but the smallest individuals of one race (grown under proper conditions) are taller than the largest of the other, and each race breeds true. Similar tall and short races exist in the sweet-pea (fig. 7), the short race being called 'Cupid' sweet-peas. When the two races are crossed—and reciprocal crosses give identical results—the offspring are not intermediate but all are tall, perhaps taller than the tall parents. When now these hybrid talls are self-fertilised, among the plants produced some are tall and others short, but again none are intermediate. Mendel regarded the tallness or shortness as distinct alternative characters, and since tallness alone appears in the first cross, he spoke of it as 'dominant,' and the shortness, which disappeared when crossed with the

3 Dwarfs 5 Talls

Fig. 7. Eight plants in the second (F_2) generation from the cross tall sweet-pea × dwarf (Cupid). The five talls and three dwarfs came from one pod of seed. (From Bateson.)

dominant tallness, he called 'recessive. More recent
work has indicated that a dominant character pos-
sesses some factor which is absent in its recessive
alternative ; in the present example the stem has
the power of continued growth which is absent in
the short pea. Dominance and recessiveness may
thus be regarded as presence and absence respectively
of the factor in question ; but since the presence or
absence of the factor may often give rise to the
appearance of an alternative pair of characters,
such a pair have been named by Bateson a pair of
'allelomorphs.' When a tall pea is crossed with
a short, the factor tallness is introduced from the
tall parent, and thus all the offspring are tall. These
are called the first filial generation, or more shortly
the generation F_1. When these hybrid (F_1) talls are
self-fertilised, their offspring (second filial or F_2
generation) consist of talls and shorts. Now it has
been seen that if the factor tallness is present it
makes itself visible, and therefore the short peas in
F_2 should contain no tall factor. And in fact when
self-fertilised, or fertilised with the original short
stock, they give only short offspring for as many
generations as the experiment has been carried to.
The tall factor has thus apparently been completely
eliminated from these short peas.

Further, Mendel found that among the talls in the
F_2 generation, some breed true to tallness when self-

fertilised, while others again give a mixture of talls and shorts. The whole result may be clearer in symbolic form. If T stands for the tall factor, t for its absence (shortness), the following results appear. (The '$2Tt$' in F_2 will be explained immediately.)

Original parents		$T \times t$	
F_1		Tt	
F_2	TT	$2Tt$	tt
F_3	TT	TT $2Tt$ tt	tt

It is thus clear that among the offspring of the F_1 (hybrid) generation, some (tt) have eliminated the tall factor altogether and show no difference from their short ancestors ; others (TT) have nothing but the tall factor and thus breed true to tallness ; and a third group, which Mendel found was twice as numerous as either of the others (therefore marked $2Tt$), proved, by giving mixed offspring when selfed, that it is hybrid like its F_1 parent.

The explanation offered by Mendel of these facts was as follows. The original tall plant produces germ-cells ('gametes') bearing tallness ; the short plant produces gametes bearing shortness (absence of tallness). The F_1 (hybrid) thus contains both conditions ; its cells, resulting from the union of two gametes, may be regarded as double structures, containing a double set of determinants for the various characters of the plant, one determinant of each pair being derived from the male parent, the other from the female. An individual produced by union of two germ-cells (gametes) and having this double character is called a 'zygote.' The F_1 zygote thus contains a determinant for tallness derived from one parent, and a corresponding determinant in which the tall factor is absent derived from the second parent. Now Mendel's hypothesis to account for the observed facts was that although the zygote produced by union of tall-bearing and short-bearing gametes contains both factors, yet when this hybrid zygote gives rise to gametes, it produces some bearing tallness and others bearing shortness, but none bearing both determinants; i.e. that the alternative characters segregate from each other in the formation of the gametes, and that gametes bearing one or other of the two conditions are formed in equal numbers. Since large numbers of gametes of each kind are formed, and since they meet indiscriminately in fertilisation, a tall will equally often meet a tall or a short, and a short will equally often meet a tall and a short, and the combinations will thus be in the ratio of $1TT$, $1Tt$, $1tT$, $1tt$, or $1TT$, $2Tt$, $1tt$. If this hypothesis is true, it can be tested by fertilising the F_1 hybrid zygote with the pure parental types; the F_1 zygote produces equal numbers of T and t gametes, the pure short race produces only t, so the offspring of the hybrid and the original short should give equal numbers of hybrid talls and pure shorts. Similarly the hybrid zygote crossed with the pure tall should give equal numbers of pure talls (TT) and hybrid talls (Tt). Mendel found that this expectation was in fact verified by experiment. The whole series may be made clearer by a diagram (p. 60), in which the zygotes are represented by squares, the gametes by circles.

The middle part of this diagram represents the production of the F_1 zygote and its offspring when self-fertilised, producing equal numbers of T and t gametes (four of each being represented) and thus giving offspring in the ratio of $1TT$, $2Tt$, $1tt$; the sides of the diagram represent the results of crossing back the F_1 zygote with the parental types TT and tt.

At this point it is necessary to explain certain convenient technical terms introduced by Bateson. It has already been mentioned that a pair of alternative characters which segregate in the gametes, as described, are called allelomorphs. When an individual is produced by two gametes bearing different allelomorphs, so that it contains both members of a pair, it is called a 'heterozygote,' or is said to be 'heterozygous' in respect of the character considered, e.g., an individual of constitution Tt is heterozygous in respect of tallness. If it contains only one kind

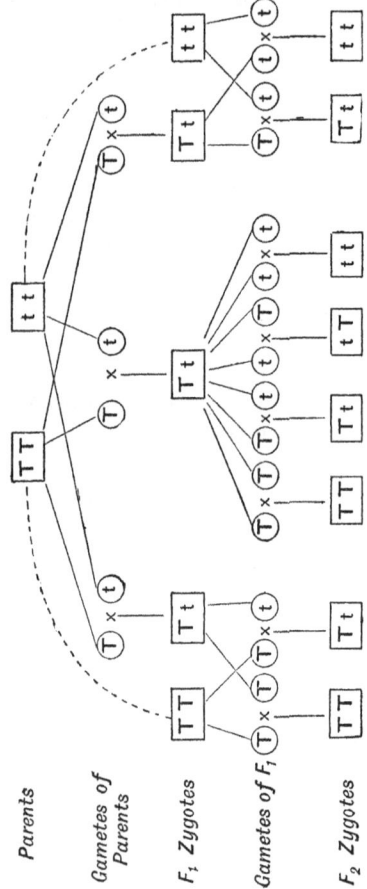

of allelomorph of a pair it is a 'homozygote,' e.g. individuals of composition TT and tt are 'homozygous' for tallness and shortness respectively. As will be seen immediately, it is possible for an individual to be heterozygous for one pair of allelomorphs and homozygous for another. The essence of Mendel's theory is that owing to the segregation of allelomorphs from each other in the production of the gametes of a heterozygote, the homozygous offspring, when self-fertilised or mated with others of like constitution, breed true to the character in question irrespective of their ancestry. As far as observation can show, the homozygous individuals TT and tt in the generation F_2 breed as true to tallness or shortness as did their pure-bred grandparents, in spite of the fact that they are the offspring of a cross.

Hitherto the original parents have been considered as differing from each other in only one pair of alternative characters (allelomorphs), but Mendel found that in the pea there were several such pairs of characters. For example, some races of peas have purple flowers, others white; these behave quite similarly to tallness and shortness. The purple flower contains a factor lacking in the white; when therefore purple is crossed with white, the purple colour is dominant and the heterozygote (F_1 hybrid) is purple. Such a heterozygous purple if self-fertilised yields 75 per cent. of purple offspring and 25 per cent.

of white ; the whites and one in every three of the
purples so produced are 'extracted' homozygotes,
being pure for whiteness or purpleness respectively,
and therefore breeding true, while the remaining
purples are heterozygous and when 'selfed' will give
both colours among their offspring.

If now a tall purple-flowered pea is crossed with
a short white-flowered, the heterozygous offspring
will be tall with purple flowers, for both these
characters are dominant. In the production of
their gametes (pollen-cells and egg-cells) segregation
will take place between tallness and shortness, and
between purpleness and whiteness, but as these pairs
of characters are totally independent of one another
they may be associated in any combination as long as
both members of a pair do not occur in the same
gamete. Gametes will thus be produced of four
kinds; if P represents purple, p its absence (white);
T tallness and t its absence (shortness), the gametes
produced by an individual heterozygous in both
characters will be PT, pT, Pt, pt, with equal numbers
of each. Since these will meet one another at random
in fertilisation, the F_2 generation will consist of
individuals (zygotes) made up of all possible combinations of these four types of gametes, viz. in the
proportion of $4PpTt, 2PpTT, 2PPTt, 1PPTT$;
$2Pptt, 1PPtt$; $2ppTt, 1ppTT$; $1pptt$.

Since purple is dominant over white and tall over

short, the first four types of zygote, which all contain both P and T, will be purple talls; the next two containing P but no T will be purple short; the two containing T but not P will be tall white, and the last with neither P nor T will be short white. The F_2 offspring will thus *appear* in the ratio of 9 purple tall, 3 purple short, 3 white tall, 1 white short. Further, of the first group one will be homozygous in both characters ($PPTT$), four homozygous in one and heterozygous in the other ($2PPTt$, $2PpTT$) and four heterozygous in both ($PpTt$). Of the remainder, one in each class will be homozygous in both characters, and the others heterozygous in one, the homozygous (pure) types being $PPtt$, $ppTT$ and $pptt$.

It is clear then that by crossing two races which differ in two allelomorphic characters, and self-fertilising (or mating together) the crossed individuals, in the F_2 generation a definite proportion of *new* pure combinations are produced. In the above example, by crossing tall purple with short white, in the second generation not only these types are produced, but also short purple and tall white, and by selecting the pure (homozygous) individuals pure races of these new types are immediately established. We thus obtain a new conception of organic characters, as factors which can be replaced by alternative characters without otherwise altering the constitution of the organism. The process is comparable with a chemical reaction, where one element may replace another in a compound ; for example, by mixing silver nitrate with sodium chloride, silver chloride and sodium nitrate are produced. Or a grosser analogy may be taken from bricks in a wall ; a red brick may be removed and replaced by a blue or a yellow one without altering the rest of the wall, and similarly in pea-plants by the process described white flowers may be replaced by purple, or yellow seed by green. After the fact of the segregation of allelomorphic characters in the production of the germ-cells of a heterozygote, the most striking result of Mendelian investigation is this discovery of the independence of characters belonging to different pairs.

That these results are not of merely academic interest is shown by the work of Prof. Biffen on wheat. Some valuable wheats are liable to the attacks of a fungus giving rise to the disease called 'rust,' other less valuable races are immune. Biffen has found that by crossing the two races together, fertilising the hybrids (F_1) among themselves, and selecting the homozygous plants in the F_2 generation, wheat can be produced which combines the valuable features of one race with the immunity to rust of the other, and so a new and most useful variety of wheat is produced. This is only one out of many examples that could be given of the possibility of combining into one race of wheat the characters previously found in different varieties.

The chief reason that breeders of plants and animals believe that the race is permanently contaminated by crossing different breeds is that commonly two breeds differ in several or many pairs of characters. If two pairs of allelomorphic

Fig. 8. A cob of Maize borne by an F_1 plant from the cross smooth (starchy) seed × wrinkled (sugary) seed, fertilised with its own pollen, showing about three smooth (dominant) to one wrinkled (recessive) seeds on the same cob. (From Bateson.)

characters are combined in the heterozygote, we have seen that only one in sixteen of its offspring is homozygous for any particular combination ; if three characters, one in 64, if four characters, one in 256, so that it is clear that Mendel's method of considering distinct characters separately must be followed, if any rules are to be arrived at for the distribution of characters among the offspring of hybrids.

Before proceeding to consider some of the further applications of Mendelian inheritance, a few examples will be given of characters in animals and plants which are found to be inherited according to this law.

In plants, flower-colour, seed-colour (due to either seed-coat or the contained embryo); production of starch or sugar in seeds (maize, see fig. 8 in which both forms of seed are shown on the same cob); hairiness or smoothness (stocks, Lychnis, etc.); 'bearded' or 'beardless' ears (wheat); 'palm-leaf' or 'fern-leaf' (Primula); long or short styles ('pin-eye' and 'thrum-eye' of Primula); pollen-shape, and also fertility or sterility of anthers (sweet-pea). Many other examples could be given; it should be noted that several of these normally occur in nature, e.g. the two flower-types of the primrose.

In animals, coloured coat and albino (many mammals); and many other colour-characters in mammals and birds; normal and long or 'Angora' hair in rabbit, guinea-pig, etc. (some doubt as to completeness of segregation); comb-characters in fowls; leg-feathering in pigeons; horned and horn-less condition (sheep and cattle); colour-characters in moths, beetles, and snails. In man, several abnormal conditions, and presence or absence of brown pigment in the iris of the eye.

As in plants, several of these cases are not in any way connected with domestication, and the wide diversity of species and characters in which Mendelian inheritance has been discovered shows that the phenomena are not rare or exceptional, but universally distributed.

It has been mentioned that of a pair of allelomorphic characters, one is regarded as generally containing some factor absent from the other, and it may be well to give an example of the kind of evidence that leads to this conclusion. In fowls there are three chief forms of comb; 'single' with a median serrated ridge, 'rose' with a broad upper surface covered with papillae, and 'pea' with a shape consisting essentially of three parallel low ridges. Rose and pea each behave as dominants to single, but when rose is crossed with pea a fourth type, 'walnut' results, which in the adult is swollen and dimpled, and, in the young at least, is crossed by a transverse band of bristles. In the Malay breed such 'walnut' combs breed true, but when made by crossing 'rose' by 'pea,' and mated together, the resulting chicks appear in the ratio of 9 walnut, 3 rose, 3 pea, 1 *single*. The appearance of singles in the F_2 generation from pure rose by pure pea is explained by the 'presence and absence' hypothesis. Rose (R) and pea (P) are each allelomorphic with their absence (r, p). A rose-combed bird is thus Rp, and a pea-combed rP,

5—2

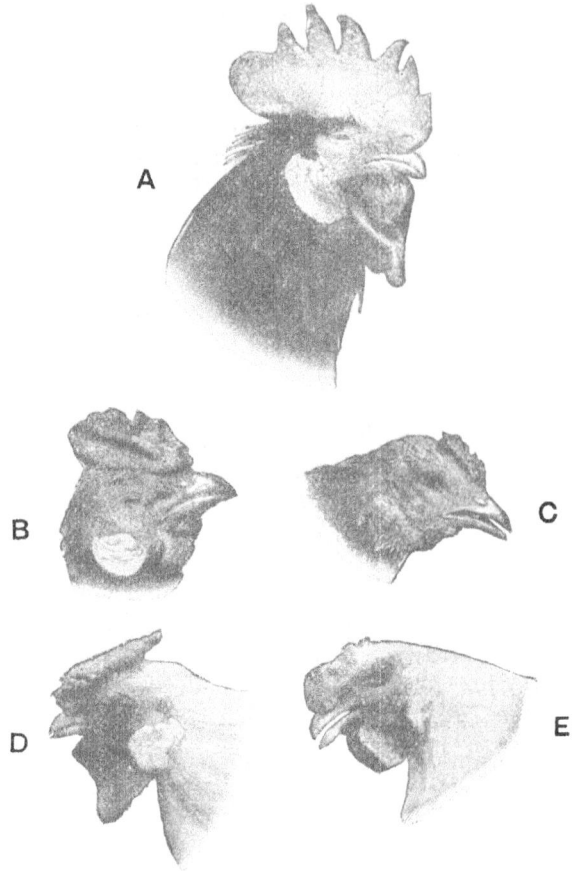

Fig. 9. Types of combs in Fowls. A. Single Comb (cock). B. Pea
Comb (cock). C. Pea Comb (hen). D. Rose Comb (cock)
E. Walnut Comb (young cock). (From Bateson.)

and the walnut combs produced by crossing them have constitution $RrPp$. They produce four kinds of germ-cells, RP, Rp, rP, rp, giving the normal ratio in F_2 of 9 birds containing R and P, 3 with R and p, 3 with r and P, 1 rp. This rp, containing neither rose nor pea is *single*, which may be regarded as the normal comb with no other factor superposed upon it.

Although the members of an allelomorphic pair usually differ from each other in that one contains a factor lacking in the other there is evidence that this is not always true, for a number of instances are known in which three or more characters are all allelomorphic with one another. In these cases it is evident that the simple "presence and absence" hypothesis is insufficient, and it must be supposed that allelomorphism exists between two or more positive characters. When two such characters are crossed together it often happens that neither is dominant, and the crossed offspring either combine the features of both parents or are intermediate between them. When these intermediate heterozygotes are mated together or self-fertilised both the homozygous parental types are produced in addition to the heterozygous form, as in the offspring of a heterozygous tall pea there occur homozygous talls and shorts in addition to heterozygous talls. The classical example of this condition is the blue Andalusian fowl. This breed

cannot be bred true; when blues are paired together
about half the chickens are blues and the remainder
evenly divided between blacks and dirty-whites. By
many generations of selection breeders have tried
without success to eliminate these black and white
'wasters,' but it remained for Bateson and Punnett
to show that if a black and a white are paired together,
only blues are produced. The two homozygotes are
black and white respectively; when these are paired
together the single black factor introduced from one
parent is sufficient to cause the crossed chicks to be
full black, and a dilute black or 'blue' results. Such
incomplete dominance, in which a single factor introduced from one parent is insufficient to bring about
the same effect in the heterozygote that is produced
by the 'double dose' present in the homozygote, has
been observed in a number of cases, some of which
must be referred to later.

CHAPTER VI

MENDELIAN HEREDITY (*Continued*)

The Inheritance of Colour

IN the simple Mendelian cases discussed in the last chapter the separate allelomorphic pairs were described as wholly independent of one another, and in the manner of their inheritance this description is correct for allelomorphic pairs in general except in special cases, of which examples will be given later. But although allelomorphs of distinct pairs are inherited independently, yet not infrequently they may react upon one another so as to give an apparently combined effect in the individual bearing them. This is especially, but by no means exclusively, seen in the colour-characters of animals and plants. In the list of examples of Mendelian characters it was mentioned that coloured coat in animals or coloured flowers in plants behave as an alternative to whiteness (albinism, i.e. the absence of pigment). But further analysis shows that the appearance of colour depends upon the presence of at least two

factors, in the absence of either of which no colour
is produced. An actual example will make this
clearer. A white rat is mated with a wild (brown
or 'grey') rat, and since colour dominates over its
absence the F_1 heterozygotes are all grey, like wild
rats. These grey heterozygotes mated together give

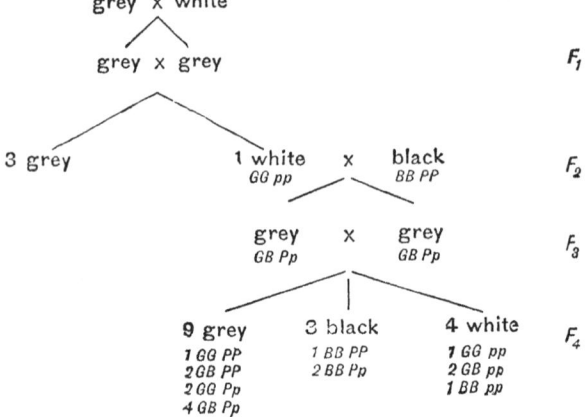

coloured and albino in the ratio of three to one. If
now one of these extracted albinos is mated with
a black rat, the offspring may not be black but *grey*,
and such grey individuals paired together will give
young in the ratio of 9 grey, 3 black, 4 white.

The explanation is as follows. For the production
of colour, two factors must be present, one for the

production of pigment in general (*P*) and the other for the determination of the actual colour of that pigment (*G* = grey, *B* = black). Neither *G* nor *B* can produce any visible effect in the absence of *P* ; a rat without *P* (represented by *p*) is thus an albino. The extracted albinos in F_2 from the cross wild grey × albino then contain *G* derived from their wild grandparent. These mated with black give *grey* offspring because grey is dominant[1] over black, and the black individual introduces the factor *P* which was absent in the albino. These grey rats (generation F_3 in the diagram) are thus heterozygous in the pair of factors grey and black (*G* and *B*) and in the factors presence and absence of *P* (*P* and *p*). They will thus produce gametes *GP, Gp, BP, Bp*, which in meeting at random will give 9 zygotes containing *G* and *P*, 3 containing *B* and *P*, 3 containing *G* and *p*, 1 containing *B* and *p*. But the combinations *Gp* and *Bp*, not having *P*, are albinos, and so we get 9 greys, 3 blacks, 4 whites.

If in the example just given nothing were known of the origin of the white rat which was crossed with the black (in the generation marked F_2), it would be said that a white variety crossed with a black had

---

[1] This explanation has been simplified by the omission of the fact that *G* and *B* do not represent factors for separate pigments, but that *G* consists in the addition of a pigment to hairs already containing *B*. A character dominant in this way is called 'epistatic,' see below p. 75.

produced 'reversion on crossing' and the young had reverted to the ancestral wild form. It is not of course necessary that the albino used to produce such a 'reversion' should itself be the offspring of a grey; such grey-bearing albinos may be bred together for an indefinite number of generations, and still carry the factor G; or if they were originally derived from a black stock they would bear the factor B. When such stocks are crossed together heterozygous GB albinos are produced, and G and B segregate from one another in the albino just as in the coloured rats in which the colour-factor P is present. The fact that colour in animals and plants depends on the concurrent action of distinct factors thus explains the phenomena of 'reversion on crossing' which have so long been a puzzle to biologists.

Among the varieties of the brown (grey) rat only two colour types occur, grey (wild-colour) and black, but in the rabbit, mouse and other animals more are found. In the mouse there are four fundamental colour-types, yellow, grey, black and chocolate. Of these yellow is dominant over any of the others, but has the peculiarity that it cannot exist in the homozygous condition; if a mouse embryo receives the factor for yellow from both parents it dies at an early stage of embryonic development, so that all yellow mice are heterozygous. Grey crossed with either black or chocolate gives grey; black with chocolate gives black, and chocolate can only appear in the absence of all the others. It seems, then, that yellow is dominant over all the other colours; grey over black and chocolate; and black over chocolate; and examination of the hair shows that in yellow and grey mice yellow, black and chocolate pigments are all present in the hairs[1]; in yellow mice the yellow pigment is evenly distributed, while in grey ('agouti') it is restricted to certain parts of the hair. In black mice the hairs contain black and chocolate pigment, but the black obscures the chocolate, and in chocolate mice this pigment alone is present. In each case, therefore, the presence of a higher member of the series obscures the lower. This is expressed by saying that black is 'epistatic' over chocolate, grey over black and chocolate, and yellow over the other three. Further, although each higher member of the series contains something that is absent from those below it, it does not seem possible to arrange the colours as pairs of allelomorphs in each of which the dominant character consists in the presence of something which is absent from the corresponding recessive; each of the colours appears to be allelomorphic with all the others, and they may best be regarded as a series of 'multiple allelomorphs' such as were mentioned in the last paragraph of the preceding chapter.

The object of this rather special digression is to

[1] Yellows heterozygous for black or chocolate only have only either black or chocolate pigment in addition to yellow.

show how the hypothesis of a series of colour-factors acting together can completely coordinate the phenomena of colour-inheritance, which very few years ago seemed hopelessly confused and subject to no definite rules. It is now possible to forecast with accuracy the results of a pairing between individuals of different colours, if the constitution of the parents with respect to the colour-factors carried by them is known. Some of these cases have been exceedingly difficult to elucidate because it is often impossible by inspection to determine the constitution of a given individual. This must be tested by suitable matings with individuals of colour lower in the series, and it is then found that the results observed agree closely with expectation.

A more surprising instance of 'reversion on crossing' was discovered by Bateson in sweet-peas. He found that within the white variety known as 'Emily Henderson' two distinct types exist, indistinguishable in appearance, which when crossed together give a purple closely resembling the wild sweet-pea of Southern Europe. The purple reversionary form in the first cross, (F_1), self-fertilised, gives in the next generation, (F_2), 9 coloured to 7 whites. The explanation is that some plants of the white form lack one colour factor (called by Bateson 'C'); others lack the complementary factor 'R,' which if present with C, would produce red pigment. Since colour can only appear when both C and R are present, each parental form is white, but when crossed together C and R are combined in one plant and coloured flowers result. The allelomorphic pairs are C and its absence (c), and R and its absence (r); the purple heterozygote is thus $CcRr$, and produces four kinds of gametes CR, Cr, cR, cr. These mating at random give offspring in the ratio of 9 with C and R, 3 with c and R, 3 with C and r, 1 with c and r. But only those containing both C and R can produce colour and therefore 9 coloured appear to 7 white. Further, among the coloured individuals of F_2, both purple and *red* appear, because the factors C and R together produce only red; to get purple a third factor for blue (B) must also be present, which can only take effect in the presence of both C and R. Since B was introduced by one only of the original whites, the F_1 purples were heterozygous for blue as well as for C and R (with fully represented constitution $CcRrBb$), and hence among the F_2 plants one quarter contain no B and in the presence of C and R are red[1].

[1] In this account, the production of colour (red) is described as being due to two factors (C and R). The recent work of Miss Wheldale [42] on the chemical nature of flower-colours indicates that the essential bodies are an organic base or ' chromogen ' and an oxidising ferment. The work of Chodat and Bach, however, indicates that such oxidising ferments must contain two components, neither of which alone is able to oxidise the chromogen and produce the coloured derivative—anthocyanin. Both kinds of white sweet-pea contain the chromogen, but it seems probable that one component of the oxidising ferment is present only in one, and the other component only in the other. Hence no colour can be produced in either. But on mating the two whites together, the mechanism for the oxidation of the

In the account given above of the colour-factors in the sweet-pea it has been shown that at least two separate elements are required to produce colour (in this case red), and a third if blue is to be present in addition. But for the production of the various shades or distribution of colour further factors are known, e.g. for the intensification or dilution of colour, and for making the wing-petals of the same or different colour from the standard. Similar phenomena are concerned with colour in animals, of which domestic varieties of the rat provide a simple instance. Rats, other than albinos, are in general either 'self-coloured,' with little or no white (this, if present, is confined to the ventral surface), or 'hooded,' i.e. white with

chromogen is again complete, and red colour (anthocyanin) is formed. The purple colour (represented by the additional 'factor' B) is due to a further stage of the oxidation of the chromogen than when only red is produced. In some white flowers (snapdragon) experiment shows that the chromogen itself may be absent.

As the purple colour in sweet-peas is due to more complete oxidation of a chromogen than red, so in animals colour-physiologists find that the series yellow, brown, black, may represent successive oxidation-stages of the same chromogen by the same ferment. The various colours of mice, for instance, are not therefore to be regarded as necessarily produced by different ferments, but the inherited ' colour-factors' determine to what stage the oxidation of the chromogen shall be carried. Some confusion has arisen from the assumption that the 'factors' postulated by students of heredity are actual specific colour-ferments, while they may be rather determinants which cause the oxidation of the chromogen to proceed to a particular stage, and may be compared with the factors which determine the production of a rose-comb or single-comb in fowls.

coloured head and shoulders and a coloured stripe along the spine. The self and hooded factors are an allelomorphic pair independent of colour, so that a hooded rat may be black or grey. The factors may also be borne by albinos, and when very young an albino bearing the factor for the hood may be distinguished by the different texture of the hair on the head and shoulders, giving the appearance of a water-mark or 'ghost-hood.' The heterozygote between self-coloured and hooded patterns differs from either parent, being black above and white below— the so-called 'Irish' type of the fancy. Such 'Irish' rats bred together always give both self-coloured and hooded rats in addition to Irish among their progeny. A similar case in rabbits is that of the well-known Dutch marking, which seems to correspond with the hooded condition in rats. In flowers the number of such characters determining the nature and distribution of colour may be considerable, so that among the offspring of a cross between two varieties of Chinese primulas or snapdragons a very large colour-series may be produced, which on first inspection may seem a continuous series from the darkest to the palest; but careful analysis of these cases has shown that the different factors may be recognised and isolated, and the series of colours falls strictly within the rules of Mendelian inheritance when each factor is considered alone.

Hitherto in discussing the interaction of distinct pairs of factors (allelomorphs), colour alone has been considered, but cases are known where colour and a structural character are interdependent in the same way. Interesting examples of this are known in stocks and primulas. When a certain smooth-leaved cream-flowered stock is crossed with a smooth white, the F_1 plants are *purple* and *hoary*, i.e. they revert to the ancestral wild purple and hoary-leaved stock. The purple colour appears for the same reason that the two forms of white sweet-pea gave purple ; one colour-factor is introduced by the white parent and its complement by the cream[1]. But the hoariness appears because the parents contain a factor for hoariness, which can only take effect *in plants with purple flowers.* The parents are therefore smooth although they contain the hoary factor. When the F_1 hoary purples are crossed together, the F_2 generation consists of purple, white and cream-flowered plants in the expected proportions, but only the purples are hoary. Smooth-leaved purple strains do exist, but these are plants lacking the hoary factor altogether ; if it were present, it would appear whenever the flowers contain purple sap.

[1] The cream-colour is due to a quite distinct factor, and the pigment is in special bodies (plastids) in the cells of the petals. The purple colour is due to a pigment dissolved in the sap, and is independent of the cream plastid-colour.

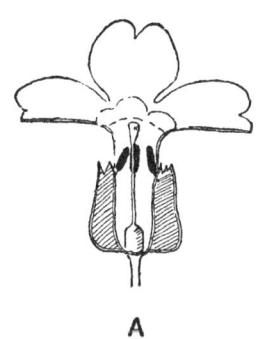

A

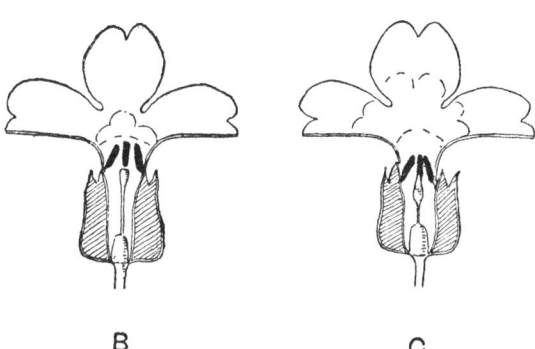

B   C

Fig. 10.   Sections of Chinese Primula Flowers.
     A.   Long-style (' pin-eye ').
     B.   Short-style ('thrum-eye').
     C.   Homostyle.

In Chinese Primulas a curious case of inter-relation between flower-colour and structure has been investigated by Bateson and Gregory. They find that the long-styled and short-styled types of flowers, so well known from Darwin's work, are an allelomorphic pair, short-styled being dominant. But when the long-styled factor is associated with a condition in which the yellow eye of the flower is enlarged to cover about half the area of the petals, the style remains short, although the anthers occupy the typical long-styled position in the tube of the flower. This condition is called 'homostyle' (fig. 10 C, p. 81). When a short-styled small-eyed plant is crossed with homostyle large-eyed, all the (F_1) offspring are short-styled and small-eyed, these characters being dominant. But in the second generation, (F_2), obtained by breeding together these F_1 plants, the following types appear:—

9 short-style with small eye,
3 short-style with large eye,
3 *long-style* with small eye,
1 homostyle with large eye.

The long-styled form has appeared in F_2 from short-style × homostyle, because homostyle is a condition of long-style modified by association with the large eye. When this association is broken, the long-style appears.

From these examples of the interaction of distinct

allelomorphic pairs, many more of which are now
known, it will be seen that many of the 'exceptions'
to the Mendelian rule which have been recorded may

Fig. 11.   Some of the types of flowers in generation F_2 from the
cross short-style (thrum) small eye × homostyle large eye.

A. Long-style, small eye. B. Homostyle, large eye. C. Short-
style, small eye. D. Short-style, large eye.

In A and B the flowers are of the 'star' type. This character
is inherited independently of the style and 'eye' characters.
(From Bateson.)

6—2

be explicable on the assumption that what appears
to be a simple character is really dependent on two
or more distinct factors, which become separated on
crossing with a different form.

In conclusion, it must be mentioned that a number
of cases are now known in which a pair of Mendelian
characters are closely associated with Sex.   In some
cases the sex of the individual determines whether
a character is dominant or recessive ; for example, if
a horned race of sheep is crossed with a hornless, the
male offspring are horned and the females hornless.
In other cases certain Mendelian characters can be
borne only by germ-cells which will give rise to one
or the other sex.   This aspect of the subject, which
includes some of the most interesting recent advances
in our knowledge of Heredity, will be considered in
a subsequent chapter (IX), after some other questions
in which the sex-factor is not directly concerned have
been discussed.

CHAPTER VII

In this chapter will be briefly considered certain questions which either are still quite unsettled, or upon which there is still active disagreement among biologists. It will be convenient to take first some which are closely connected with the Mendelian theory of heredity, and pass on later to others which are related equally to any theory of inheritance which may be adopted.

One of the chief lines of attack on the Mendelian theory has been the proposition that the absolutely complete segregation of allelomorphic characters in the germ-cells, postulated by that theory, has not been proved. If the theory is rigidly true, then in the case of a tall pea crossed with a short (Chap. v) the homozygous talls and shorts among the offspring of the cross should be as pure for tallness or shortness as the original parents; neither character should

have been influenced in any way by its association
with the other. It has been maintained that the
Mendelian categories are not sufficiently definite to
allow such a statement to be made with certainty.
The Mendelian can only reply, that in the great
majority of cases the 'extracted' pure individuals in
the F_2 generation do not differ recognisably from
the original parents in the characters considered,
and that no signs of impurity can be found in later
generations.

There are however instances in which it appears
that Mendelian segregation may not be perfect. It
has been maintained that an instance of this is
provided by hair-length in guinea-pigs. When a
long-haired ('Angora') guinea-pig is mated with
a short-haired, the F_1 offspring are short-haired,
shortness being dominant, owing perhaps to the
presence of a factor which prevents the growth of
the hair after reaching a certain length. But when
such F_1 (heterozygous) short-hairs were mated
together, in addition to apparently pure longs and
shorts, animals with hair of intermediate length
were produced, and these crossed back with pure
long-hairs gave no short-haired young. It is suggested
that the long and short characters have become fused
in some germ-cells, segregation being incomplete or
non-existent, so that germ-cells bearing the mixed
character are produced. Again, in a cross between

lop-eared and short-eared rabbits, young with ears of intermediate length are produced, and these mated together give no evidence of segregation in the next generation. From these and some other similar observations it must be concluded, either that in some cases there is incomplete segregation or even complete fusion of alternative characters, or more probably that what appear to be simple characters are really complex, and that the true-breeding inter-mediates are formed by a new combination of ele-mentary factors. An instance, which is perhaps similar, will be mentioned in the next chapter in discussing the inheritance of pigment in Man.

A second question with regard to Mendelian segre-gation concerns the inheritance of such apparently 'continuous' variations as were illustrated by the example of Johannsen's beans (Chap. IV). It was seen that each 'pure line,' derived by self-fertilisation from an individual plant, has its own type about which the size of the beans borne by the plant fluctuates. In a number of cases of this kind, especially where races differ in such characters as size, it has been found that the size depends on two or more independent factors, each having a similar effect, and each inherited according to Mendel's law. It seems probable, there-fore, that the differences between Johannsen's 'pure lines' may depend on the number of such size factors present in the several lines. When several factors cooperate thus in producing a character in such a way that each increases the total effect, they are spoken of as 'multiple factors.'

Another subject which Mendelian investigation has brought into prominent notice, and which has led to much controversy, is the kind of variation which has been effective in the process of evolution. Darwin assumed that evolution takes place by the preservation of very small 'continuous' variations which occur in a direction favourable to the species, but even among his immediate followers, for example Huxley, doubt was expressed whether larger step-like variations or 'mutations' may not have been operative. Darwin rejected this idea chiefly on the ground of the rarity of such mutations, which makes it inevitable that the mutating individual should generally mate with one of the normal type, and so it was supposed that the mutation would be diluted and rapidly lost. But Mendelian work shows that this dilution does not occur in simple cases; the offspring of the cross between the mutation and the type produces half its germ-cells bearing the mutation to its full extent, and these will transmit the mutation until the race may become widely infected with it, and not infrequently individuals both of which possess it will mate together. A dominant mutation which has come into existence and spread widely in this country during the past seventy years is the well-known black variety of the 'Peppered Moth' (*Amphidasys betularia*), and a recessive mutation would equally often be represented in the germ-cells of many individuals and would appear when two which bear it mate together. Neither a dominant nor a recessive mutation will of itself increase in frequency beyond a certain point, for the population, if mating be at random, will soon reach a condition of stability; if, however, the mutation be advantageous so as to be preserved at the expense of the type by natural selection, it will increase rapidly until it replaces the original form. But the difficulty has naturally been felt that the marvellously perfect adaptations which are so frequent in nature cannot be imagined to have arisen by large steps, but must have been acquired gradually, and therefore many naturalists reject the suggestion that mutations can have been largely operative in evolution. The fallacy here is the assumption that all discontinuous variations must be large; the case of Johannsen's beans shows that essentially stable variations occur, which probably differ from mutations only in their small extent, and by the selection of such 'minute mutations' the wonderfully perfect structures of living things might be produced. It may perhaps be regarded as hair-splitting to distinguish between minute mutations and fluctuating variability, but the distinction lies in the nature of their inheritance, which is the essential thing in evolution. It has been seen that no selection within the 'pure line' in the case of the beans has any effect; for progress to be made, a new mutation, small though it might be, is necessary.

The problem in heredity which has probably
given rise to more controversy than any other is that
alluded to more than once previously, of the inheritance or non-inheritance of acquired characters, that
is, characters produced in the individual during its
life by the action of some sort of stimulus. Some
aspects of the question have already been considered,
and from what has been shown of the very definite
nature of the inheritance of germinal (inborn) characters, it will be understood why students of heredity
are increasingly disposed, *a priori*, to disbelieve in
the transmission of acquirements; for if these were
transmitted to any considerable extent, this fact
must interfere, one would suppose, with the orderly
appearance in the offspring of the characters represented in the germ-cells of the parents. But at the
present time no treatment of heredity could be regarded as complete without some mention of the evidence which has been adduced in favour of the
transmission of such characters. Unfortunately, the
evidence is almost always capable of interpretation
in more than one sense. The supporters of the belief
in transmission rely largely on indirect evidence,
especially on the difficulty of imagining any cause
of evolution in certain directions if the effects of
acquirement are excluded. A vast literature has
grown up around this question, of which only
illustrative examples can be given. In animals which

live in the dark the pigment in the skin is frequently
absent, as it is also in flat-fish on the side of the body
which lies protected from light on the sea-floor. It is
said that pigment in such cases cannot be harmful, and
so its disappearance is not due to natural selection.
But pigment very generally appears in response to
the action of light, and so it is supposed that the
absence of the stimulus to production, acting for
many generations, has caused the pigment to dis-
appear. This is illustrated by the well-known
experiment of Cunningham on flat-fish. The young
fish is pigmented on both sides of the body; it
then settles on one side and the pigment on that
side disappears. Cunningham reared such young
fish in an aquarium lighted from below: when they
settled on the bottom the pigment disappeared, but
if kept still longer exposed to light from beneath,
the pigment began to come back again. The disappearance of the pigment, although exposed to
light, proves that the loss is hereditary; its return
on continued exposure to light is interpreted by
Cunningham to mean that its disappearance was
due to absence of light, and has gradually become
hereditary, but that the process can be reversed by
again exposing the lower side to the action of the
stimulus. The same argument has been used with
regard to the colourless skins and vestigial eyes of
animals living in caves; where the structure is use-

less it disappears, and the most obvious cause to
assume is lack of use, which, acting cumulatively
through many generations, has become hereditary.
Such evidence, however, is only presumptive, it does
not amount to proof, and on the other side may be
adduced the pigments of birds' eggs. Birds which
nest in holes or dark places usually have colourless
or slightly coloured eggs, while those which lay in
open places have eggs more or less matching their
surroundings. This appears closely comparable with
the condition of skin-colour in fishes and amphibians,
and yet it is impossible that the action of light could
have any direct effect on the production of pigment
in birds' eggs. That the loss of pigment in each case
is connected with its uselessness is probable, but the
birds' egg case seems to show that it is not due to
' use-inheritance.'

A second instance of the indirect evidence for the
inheritance of acquired characters may be given, that
of instinct. Instincts are very similar to firmly rooted
habits, and have been regarded as habits which from
being performed through many generations have
become hereditary. There can be no doubt that, in
the higher animals especially, instincts may be rein-
forced and perfected by habit, but many cases can be
adduced in which it seems impossible that habit has
played a part in the evolution of an instinct. Many
insects have exceedingly perfect and complex in-

stincts in connexion with egg-laying, yet the process may last only a few hours, and the eggs may all be fully formed and ready for laying before the insect hatches from the pupa. In the worker-bee, too, there are many admirably developed instincts, and also structural features which might be thought to have originated by the transmission of acquired adaptations, and yet the worker-bee, except in rare cases, never reproduces itself, but is produced by a queen and a drone with structures and instincts different from its own. If in these cases we find perfect instincts which cannot have arisen by the inheritance of acquirements, it seems unreasonable to assume that instincts in other species must have arisen in this way. These two cases are given merely as examples of the presumptive evidence that has been brought forward. It is admitted that the process of evolution would be more easily comprehensible if the inheritance of acquired characters were a fact, but it is clear that no absolute proof of its existence can be based on cases of this kind.

Exact experiments on the possible inherited effects of acquirements are difficult to devise so as to be unequivocal, and most have given negative results[1]. A case which at first sight seems to prove

[1] Experiments on moths and butterflies have been mentioned in Chapter III; in some, notably those of Fischer with the Tiger Moth,

the inherited effects of conditions is the experiment
of Kellogg in starving silkworms, in which he found
that when the caterpillars were starved for two
generations, the third generation, even if well fed,
were below the normal size. But there is here a
possible source of error, that the eggs produced by
starved females may have been lacking in yolk, so
that the resulting caterpillars would be weakly from
the beginning and never overtake the normal size.
If so, the apparent effect of inheritance of bad
conditions would be due really to poor embryonic
nourishment, not to germinal difference. The same
explanation might apply to the apparent cumulative
effects of under-feeding in man, if the mother cannot
adequately nourish the infant before birth. The
famous experiments of Brown-Séquard on the inheritance of artificial injuries in guinea-pigs must
be mentioned. He found that when the parents
were subjected to operations of various kinds, some
of the young showed corresponding abnormalities,
especially in the case of the effects of certain injuries
to the nervous system. Subsequent experiments
however have not completely confirmed his results,
and there is reason to believe that where they have

Arctia caja, definite evidence for the transmission of modifications
was obtained, but this may have been due to direct modification of
the germ-cells themselves.

been confirmed there are other possible explanations of the apparent transmission of the effects of injury. For example, Brown-Séquard found that when the chief nerve of the leg is severed, the toes become morbid and the animals frequently nibble them away. A small percentage of the offspring of guinea-pigs lacking toes from this cause also had toes missing. But it has been pointed out that rodents in captivity sometimes eat off the toes or tails of their young, and if the mother had acquired the habit of nibbling her own toes, she might bite off those of her young shortly after birth and give the appearance of the inheritance of a mutilation [20, 29]. Evidence has also been brought forward that the mutilation of the parents may cause the production of a toxin, which is transmitted directly to the offspring, and causes abnormalities to appear in them in the same organs that were injured in the parent.

Kammerer [19] in Vienna has made some remarkable observations on salamanders and a species of toad which seem to support the idea of the inheritance of acquired characters. For example, among other experiments, he finds that the animals can be accustomed to lay their eggs in water instead of on land, and the young become modified to suit their new surroundings, and the modifications are progressively increased in later generations. He points out however that most of his results, like those obtained by

Tower (Chap. III), may have been brought about by
the action of environment on the eggs at the time
of maturation, but they differ from Tower's in the
regularity of their appearance and in being adaptive.
It should be said that doubt has been cast upon the
accuracy of both Tower's and Kammerer's results.
Reference must also be made to the work of Sumner
[32 a] who finds that rats kept at a high temperature
differ in several particulars (proportions of the body,
etc.) from those kept in a cold room. These characteristics are inherited to a measurable extent, and the
explanation offered in the case of insects, that the
germ-cells were directly affected by the temperature,
is not easy to apply to a warm-blooded animal. There
may of course be indirect effects, brought about by
a change in general constitution, but the admission
of this hardly differs from accepting the inheritance
of acquired characters. A considerable list of cases
of the transmission of environmental modification,
chiefly but not entirely in plants, is given by M°Dougall
[22 a].

The question of the inheritance of acquired
characters is complicated by the fact, already referred
to in Chapter III, that external stimuli may act not
only on the body itself, but also on the germ-cells, and
that in consequence changes may appear in the offspring which are not in any true sense due to inheritance from similar changes in the parent. A remarkable

example of this sort is provided by the observations of Agar on the water-flea *Simocephalus*. He found that when this animal was fed on certain microscopic plants, the valves of the carapace ('shell'), instead of bending towards each other like the shells of a cockle, became reflexed or bent outwards at their lower edges. When the specimens with reflexed shells were removed to normal conditions a few hours before their first eggs were laid, the young from these eggs, although reared under normal conditions, developed reflexed shells. The second and third broods of young, also produced under normal conditions, still showed the effect in a diminishing degree, and the abnormality is found also to a much slighter extent in the next generation (grandchildren of the individuals which originally acquired the character). Later generations did not show it. Agar believes that in consequence of being fed on abnormal food, the water-fleas produce a substance which stimulates the shells to bend outwards, and that enough of this substance is present in the eggs of abnormal individuals, even though they are laid under normal conditions, to induce the abnormality in the offspring, and even in the second generation. Evidently, then, if Agar is right, the case is not properly one of transmission of an acquired character, but of a temporary modification of the germ-cells of such a kind as to induce an abnormality like that present in the parent.

An example in some ways comparable, but differing in one important respect, is provided by Stockard's experiments on treating guinea-pigs with alcohol. Stockard subjected guinea-pigs each day to the vapour of alcohol, until they became thoroughly intoxicated, and investigated the effects on the offspring. The treated animals themselves remained healthy, and even after almost daily intoxication for three years, showed little sign of harmful effects, except a decrease in fertility. Among the offspring, however, there was a large proportion of defective individuals, and when treated males were mated with normal females a higher proportion of the young died early than among the young of treated females mated to normal males. When the offspring of two treated parents were mated together, their children and grandchildren were still more defective. The defects, in addition to diminished fertility, were chiefly abnormalities of the eyes and nervous system—blindness or even complete absence of eyes on one or both sides, and partial paralysis. In this case, there is no question of the transmission of an acquired character, for the alcoholized parents were scarcely affected; the results appear to be due to injury inflicted on the germ-cells, resulting in abnormalities in later generations. It must not be assumed, however, that Stockard's observations can be applied indiscriminately to other cases, for similar experiments with mice, fowls and moths have given different results. In all these animals it has been found that the offspring of alcoholized parents, instead of being defective, are actually stronger than those of the normally treated controls. Pearl [23 a], whose experiments with fowls are the most extensive, explains this apparent contradiction by supposing that in guinea-pigs the germ-cells are injured but not destroyed by alcohol, but that in fowls, if the alcohol has any effect on the germ-cells, it destroys them, and only leaves the most resistant alive. These stronger germ-cells would thus be selected as the source of the next generation, and hence the offspring of alcoholized parents, in which the selection has occurred, would be on the average stronger than those of normal parents.

On the whole, the hypothesis of the inheritance of acquired characters must be regarded as 'not proven,' and our increasing knowledge of the definiteness of many germinal characters makes it doubtful whether it can be a factor of great importance in the constitution of the individual, or to the course of evolution. Some further evidence in this direction will be given in the next chapter[1]. It is at least quite

---

[1] The recent evidence which has been brought forward on the subject warns us against a dogmatic denial of the possibility of the inheritance of acquired modifications. The number of cases recorded is now considerable, in which adaption to changed environment, either of structure or instinct, appears to be transmitted to the next generation. The tendency of biological thought is certainly towards

certain that some environmental effects are not inherited, even when lasting over many generations. An interesting instance of this has been given by Morgan. In his experiments with the fruit-fly, *Drosophila,* a mutation appeared in which the dark bands on the abdomen were irregularly reduced or absent. It was found that this character only shows itself when the flies have been reared in damp surroundings; flies of the same brood reared in dry bottles have normal pigmentation. It is thus possible, by breeding the flies in dry environment to obtain for several generations nothing but insects quite normal in appearance, but by rearing the offspring of these with the required amount of moisture, flies as abnormal as the original abnormal stock are immediately produced.

a recognition of the unity of the organism as a whole, including its germ-cells, and especially where the organism adapts itself to change, it seems possible that this adaptation is transmissible. The belief that 'somatic' changes could not be transmitted rests largely on the idea that every character is determined by a 'factor' or determinant in the germ-cell, but it is clear that any character is not developed directly from the germinal determinant, but by the relation existing between the determinant and its surroundings, viz. the body of the organism. If the surroundings are changed, this relation may be altered, and the altered relation may be transmitted to the offspring, so bringing about a corresponding change in the character as it appears in the next generation. How far such changes are of real importance, and whether they are permanent or, so to speak, temporary expedients to meet changed conditions, is still an open question. Students of heredity generally would take the latter view.

A few minor questions remain. One of these, which has played a considerable part in biological literature, is the alleged phenomenon called Telegony. It was formerly believed, and the belief is still firmly held by fanciers and animal breeders, that if a female of one breed bears young by a male of another breed, and is then mated with a male of her own kind, the offspring of this second mating will in some cases show the influence of the first sire, and instead of being pure-bred will in some respects be mongrels resembling the mongrel offspring of the first mating. The instance of this made classical by Darwin is 'Lord Morton's Mare,' in which a chestnut mare bore a colt by a quagga, and afterwards two colts by a black Arab stallion, both of which were dun-coloured, and bore stripes on the legs and in one colt on the neck also [7]. But it is known that dun horses are frequently striped to some extent, and Ewart's well-known work with zebras [11], in which it was attempted to repeat this experiment, gave negative results. The belief in telegony is widely held among dog-fanciers, and many cases could be quoted, but whenever properly controlled experiments have been made, no evidence of telegony has been forthcoming. The belief in it is almost certainly due to the habit of generalising from individual instances; whenever a case occurs which appears to favour the belief, it is adduced as proof, even though other causes may have

been operative, and matings in which no evidence for
it appears are passed over in silence. If it were a
genuine phenomenon, it is almost certain that conclusive evidence for it would have been obtained in
the numerous breeding experiments recorded in recent
years.

Another idea very widely held, but apparently
resting on no better evidence, is the belief in maternal
impressions, especially in the case of mankind.   By
maternal impression is meant the influencing of the
child by events affecting the mother during pregnancy·
It is commonly believed that if a pregnant woman
is injured in any part, or even sees an injury to
another person so as to excite her imagination, the
corresponding part in the child may be abnormally
developed, or may bear some mark, caused, it is supposed, by an impression conveyed from the mother.
More general still is the belief that the temperament
of the child is influenced by the mother's mental
condition during pregnancy.   This latter belief is
scarcely susceptible of accurate investigation, but
the belief in bodily marks or malformations being
due to corresponding injury to the mother, or to
her attention being strongly attracted to that part,
is almost certainly based on coincidence. A large
number of children are born with some abnormality,
and a very large proportion have some mark on the

skin. Many mothers during pregnancy undergo some
slight accident or see some deformed person, and thus
it must happen that a mark on the child will often
roughly coincide in position with the part affected
in the parent.   If every coincidence of this kind is
quoted as proof of the reality of maternal impression,
and the cases are left unheeded in which no relation
can be found between abnormality in the child and
events affecting the mother, it is natural that a belief
in the phenomenon will easily take firm root.   The
evidence available however is probably insufficient
to support any other view than that of accidental
coincidence.

CHAPTER VIII

In the chapters dealing with the various aspects of heredity in general, a number of instances have been given illustrating inheritance of various characters in man, and the province of this chapter will be to collect and add to these cases, so as to sketch the general outlines of what is known of human inheritance. It has been seen that as man differs in no important way in his bodily characters from the other mammalia, so the laws governing the variation and transmission of those characters are the same as are found throughout the animal and vegetable kingdoms wherever they have been investigated; and further that the 'mental and moral' attributes of man, which presumably are correlated with physical structures, are inherited no less strongly than the bodily features themselves. When investigated by the biometric methods, the stature, span, length of fore-arm, eye-colour, and certain other physical characters or measurements, are found to

give a parental correlation and thus an intensity of
inheritance closely similar to those obtained from
the study of various animals and plants. When
various non-measurable and less definite characters
such as intellectual ability, hand-writing, etc. are
investigated by the same methods, a similar intensity
of heredity is found, and finally the same is true when
the character chosen is liability to certain diseases,
notably tuberculosis and insanity, or such abnormal
conditions as congenital deafness. Since these latter
conditions have been only briefly alluded to, and are
of such fundamental importance for the well-being
of mankind, the evidence may be referred to rather
more fully here. The case of insanity is especially
convincing, for it is not open to the objection sometimes made with regard to infectious diseases that
the cause of the apparent inheritance of the condition
is the exposure of the child to infection from the
parent. It must be remembered that there are
many kinds of insanity, in one of which at least
(chorea), the inheritance appears to be Mendelian,
and that of two men with equal tendency to mental
aberration, one who is not exposed to strain may
remain normal through life, while another under
more arduous conditions may break down. But the
data of occurrence of insanity among tainted stocks
make it certain that 'the insane diathesis is inherited
with at least as great an intensity as any physical or

mental character in man. It forms...probably no
exception to an orderly system of inheritance in
man, whereby *on the average* about one-half of the
mean parental character, whether physical, mental or
pathological, will be found in the child. It is accord-
ingly highly probable that it is in the same manner
as other physical characters capable of selection or
elimination by unwise or prudential mating in the
course of two or three generations.' (Heron [10]).
Similarly for congenital deafness, Schuster writes
'...that striking confirmation has been obtained of
previous work on widely different characters, at any
rate with regard to the correlation between father
and children, and mother and children.' [31].

These examples, which might be added to, of
results obtained by 'biometric' methods, make it
sufficiently clear that a knowledge of the facts of
inheritance is of importance to mankind, and that
the further collection of accurate data is one of the
most needed social requirements. Before passing on
to other aspects of the question one other subject
may be mentioned. The measure of resemblance
in these characters has not only been worked out
between parent and child, but between brothers
and sisters, between children and grandparents and
uncles and aunts, and between cousins. Some estimate can therefore be made of the probability of an
individual being affected if his relatives are known,

a thing which should not only be useful to insurance offices, but to all thinking men, for it may ultimately become the basis for deciding on the propriety of marriage by members of tainted families. In general it appears that the resemblance of a child to its grandparent is rather more than half of that to its parent, and that the resemblance between uncle and nephew, or between first cousins, is very slightly less than between grandchild and grandparent.

We may now turn to definitely discontinuous characters in man, some of which are clearly Mendelian in their inheritance. One of the most interesting cases is that of eye-colour. Hurst [17] has shown that complete absence of pigment in the front of the iris is recessive to the presence of pigment; that is to say, that two pure blue-eyed people have only blue-eyed offspring, but that a blue-eyed individual married to one with any brown or yellow in the iris may have children with pigmented eyes and that two pigmented parents have pigmented children, with or without a proportion of blue-eyed in addition. Within the pigmented class there is great range of variation, from a small yellow rim round the pupil to completely dark eyes, but if the several pigmented classes are grouped together, Hurst's evidence makes it quite clear that the characters 'pigmented' and 'non-pigmented' are a Mendelian pair. This is thus a case in which the occurrence of apparently continuous variation within a discontinuous category is clearly shown. Of other Mendelian characters in man, colourblindness, complicated by its relation with sex, has already been mentioned and is further considered in

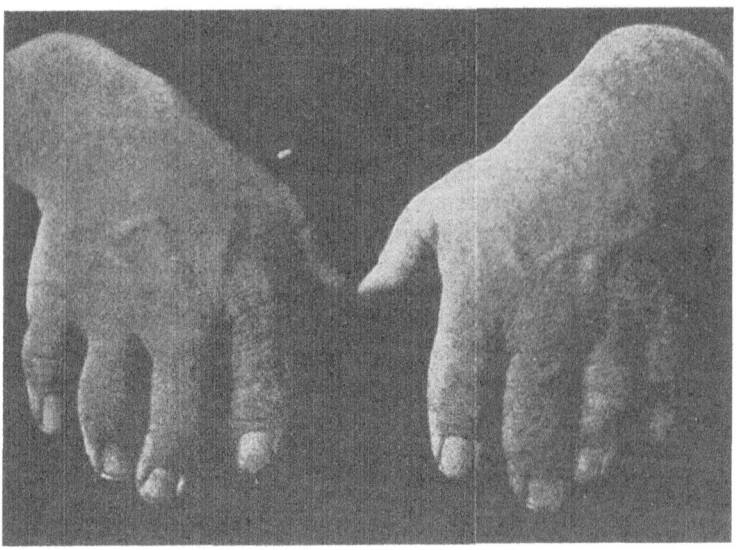

Fig. 12. Brachydactylous hands. (From Bateson, after Farabee.)

Chapter IX. Several cases are known in which an abnormality behaves as a simple dominant, e.g. the condition of the fingers known as 'brachydactyly,' in

which the fingers have one joint less than the normal;
most, but probably not all cases of congenital cataract,
and several other diseases of the eye.   Perhaps the
most remarkable human pedigree ever collected is
one of 'night-blindness,' extending through nine generations and going back to the seventeenth century,
which has been published by Nettleship (see [1]).
The condition is one in which the patient cannot see
in dull light, and it behaves as a Mendelian dominant,
probably, however, with some complication, since the
numbers affected are less than the theoretical expectation.   In all these cases in which the abnormality
is dominant, only affected individuals can transmit it;
the normal members of the family have only normal
offspring, a condition which is shortly summarised as
' once free, always free.'

The rule that the affected alone transmits will
be followed only when the condition depends on a
single factor; if it depends on more than one, or if
its dominance is modified by sex or other conditions,
then non-affected individuals may have affected offspring.   This is possibly the case in many diseases
in which it appears that the affection is dominant,
and yet certain non-affected individuals have affected
offspring, and in such examples it must also be remembered that the disease is probably not always
developed in people in whom the tendency is present;
the tendency may be there but the conditions required

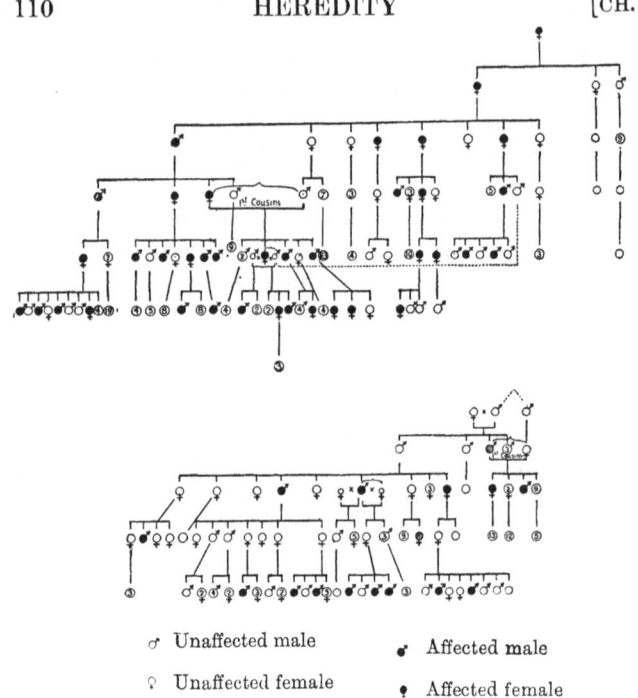

♂ Unaffected male   ♂ Affected male

♀ Unaffected female   ♀ Affected female

♀ 3 females

♂
} Not certainly affected   ⊙ 3 individuals, sex not recorded
♀

⌣⌣ Indicates consanguinity.

Pedigrees of *Retinitis pigmentosa*, from Nettleship. The upper pedigree illustrates normal transmission of a dominant character, transmitted only through affected parents; the lower pedigree shows 'sex-limited' transmission, with two (exceptional) affected females.

to bring out the disease may be avoided, especially if it is a condition not present at birth, but appearing later in life. This kind of thing may perhaps be illustrated by the pedigrees of *Retinitis pigmentosa* taken from Nettleship (Bowman Lecture, 1909 [23]) on p. 110.

The disease is not usually present at birth, but comes on at a varying age, sometimes during or after middle life, and it will be seen that in the first pedigree it is transmitted only by affected members of the family, so behaving as a typical Mendelian dominant. In the second pedigree, however, it commonly 'skips' a generation, the parents of affected individuals usually being normal, but themselves children or sisters of those who are affected. The fact that in both families (as in most cases of this disease) males are more frequently affected than females, suggests that there is some complication, and this is perhaps connected with the fact that in one family the disease behaves as a simple dominant, while in the other it is most frequently transmitted, like colour-blindness, through normal females from affected males. These pedigrees are given as examples of the somewhat irregular inheritance of diseased conditions such as may frequently be seen in the medical journals; many of them are probably explicable in the ways suggested above.

A somewhat different group of phenomena is

illustrated by the inheritance of pigmentation in man, in skin- and hair-colour. In the case of hair-colour, Hurst has given evidence that red hair is recessive to non-red, and that typical Mendelian segregation takes place. In respect of the various shades of light, brown, and black hair, however, the colours grade into one another so continuously that it is impossible to place them with confidence in Mendelian categories, and the matter is complicated by the fact that both red and brown pigments may be present together, and that the amount of brown increases from childhood to adolescence. It seems established that red only appears if the factor for it is homozygous (i.e. received from both parents), and Hurst agrees with Davenport in finding that the darker shades of brown are dominant over the lighter, with the result that, with very few exceptions, children do not have hair darker than that of their more dark-haired parent. Davenport finds a similar condition in eye-colour— that a greater amount of dark pigment on the iris is dominant over a less, and Hurst finds that the evenly pigmented iris is dominant over the ringed pattern. We thus find the pigmented eye dominant over the non-pigmented, and within the pigmented category several minor characters, each of which almost certainly shows Mendelian inheritance. The very frequent but not absolutely perfect correlation between dark eye-colour and dark hair suggests that

similar factors may perhaps be at work in both
cases[1].

The inheritance of skin-colour in man is also one
of the cases which has hitherto defied Mendelian
analysis, and has been quoted more than once as
disproving the universality of Mendelian inheritance.
When a 'white' European marries a negro, the offspring are 'mulattoes,' intermediate between the
parents. Mulattoes however are not all alike, some
have brown skins and some yellowish. When they
marry among themselves they are said never to
produce full blacks or full whites, but again mulattoes, who however vary in the depth of their colour.
When a mulatto marries a white, the 'quadroon'
offspring are lighter than mulattoes but usually
darker than Europeans; there is evidence however
that they vary considerably, with possibly a certain
amount of discontinuity between the darkest and
lightest. Some evidence of segregation is also provided by the occasional instances of 'throw back'
to very dark skin and negroid features or hair among
children of apparently white people with some negro
ancestry. Davenport, on the basis of over one hun-
dred pedigrees collected in Jamaica, Bermuda and

---

[1] In this connexion it may be mentioned that two recent writers
have maintained that the famous 'Habsburg lip' and the peculiar
Jewish physiognomy are Mendelian characters, the former dominant,
the latter recessive to the normal type.

Louisiana, concludes that there are two separately inherited factors for black pigmentation of the skin in the negro, both absent from the European, and that each of these factors has more intense effect when homozygous than when heterozygous. There is also a factor for yellow pigment independent of the two black factors. Evidently, then, there will be a considerable series of categories of skin colour among the children of two mulatto parents, giving the appearance of an almost continuous series from the darkest to the lightest. The woolly hair of the negro is also due to a Mendelian factor which must be homozygous to have its full effect, and which is inherited independently of the skin-colour factors. There is, however, a strong correlation between the colour of skin and hair. If Davenport's conclusions are substantiated, it would seem that the negro differs from the white man in respect of at least four independent factors affecting the skin and hair, in addition to factors not yet analysed for such features as the shape of the head, nose and lips, and that only when all these characters happen to be combined in one individual would a fully negro child appear among the offspring of mulatto parents. In this respect crosses between different races of mankind resemble hybrids between different species of animals or plants, except that there is usually no sterility. Most of the Mendelian investigations have been made on

varieties which differ in few characters, for the sake
of simplicity, but when species are crossed and the
offspring are fertile so many diverse characters are
concerned, of which the relation to one another is
not generally known, that the offspring of the hybrids
may contain no individuals closely resembling either
parent species. This has been explained by saying
that only varietal and not specific characters segregate from one another on the Mendelian scheme,
but it is not improbably due to the multiplicity of
characters concerned, and their complicated inter-
relations, which make analysis exceedingly difficult.
It is also not impossible, when germ-cells differing
very considerably in constitution combine in fertilisation, that in the formation of the germ-cells of the
next generation the machinery for segregation is
inadequate. Extreme cases of this are possibly the
cause of the frequent sterility of hybrids, but it may
be that when the parental differences are insufficient
to prevent the formation of fertile germ-cells, they
may yet be enough to interfere with normal Mendelian
segregation.

Certain aspects of inheritance in mankind have
now been reviewed, and it remains briefly to indicate
the lines on which our knowledge may be of practical
importance. One of the things which is especially
prominent when the evidence is considered as a
whole is the exceeding definiteness or determinancy

of the process of heredity. Given parents of certain
constitution, it can be said with confidence that *on
the average* a certain proportion of their offspring
will have such and such characters. It matters not
whether the character considered is regarded from
the standpoint of the Biometrician or the Mendelian,
both agree that what is present in the germ-cell will
be present in the individual, and that external conditions as a rule play but a small part in determining
its appearance. The Biometrician finds an average
value for the intensity of inheritance, and shows that
it is sensibly the same whether the character considered is stature, eye-colour, ability, or tendency to
congenital disease. When the character in question
is a simple case of presence or absence, the Mendelian finds that it is present in a definite proportion
of the children of affected parents, so that he can say
with confidence that among the offspring of a parent
who has congenital cataract or abnormally jointed
fingers, about one-half will be similarly affected, and
there is no hope in such a case that the severity of
the affection will diminish in later generations. Where
the disease depends on several factors, it may perhaps
be eliminated by repeated marriage with untainted
stock, but in such cases as cataract or colour-blindness there is no hope of this.

It is commonly supposed that inherited disease
arises largely from the cumulative effect of bad con-

ditions, drink and the like, but it has been seen how
doubtful it is whether the effects of such things
are really transmitted, and in any case it can be
proved that in comparison with the germinal constitution, the effects of environment are relatively
insignificant. Galton was one of the first to illustrate
this by the study of twins. Human twins are of two
sorts; in one case they arise by the simultaneous
development of two ova, as in the litters of lower
animals, and then they are no more alike than other
children of the same parents, and may be of different
sexes. Twins of the second kind are probably produced by division of one ovum, and are then of the
same sex and so alike as to be called 'identical.'
Such 'identical' twins remain through life, despite
differences of environment, more like one another
than successively born brothers commonly are, even
when brought up in precisely the same surroundings.
The same thing has been shown by an investigation of
school-children in relation to their home environment
and the habits of their parents. From a study of over
70,000 children in Glasgow, classified according to the
employment or non-employment of their mothers in
work outside the home, it was found that the relation
of their height and weight to the employment or
non-employment of the mother was almost negligible
compared with the relation between the physical
characters of the mother and child. Still more

surprising, if correct, is the observation that no
regular relation could be found between drinking
habits in the parents and the health, intelligence
or physical development of some 1400 children in
the schools of Edinburgh. [Elderton, 10[1].] Investigations of this kind are still in their infancy, and
perhaps more urgently needed than any other social
data—and it would be rash to make sweeping general
statements from the little that has been done. The
opposite results obtained by Stockard (32 *b*) and by
Pearl (23 *a*) in their experiments on the treatment
of guinea-pigs and fowls with alcohol show that the
problem is by no means a simple one, and that the
results obtained from animals cannot be applied un-
critically to man. Results like the examples quoted
make one doubt whether the generally accepted statements about the degeneration caused by unhealthy
conditions or drink are really at all reliable. It is
easy where insanity or other disease occurs, to say
that in so many per cent. of the cases there has
been alcoholism in the ancestors, and that therefore

[1] The conclusions arrived at from this investigation have been
severely criticised from both the medical and the statistical side, and
it is probable that the statistical material used is not capable of
yielding a decisive answer to the question whether drinking habits
in the parents cause deterioration of the children or not. Neverthe-
less, the fact that it was not found suggests that the habits of the
parents are relatively unimportant compared with the nature of the
stock in determining the character of the children.

alcoholism is a cause of insanity; but in the first
place it must be shown that the alcoholism is not the
result of nervous disorder, which in the next generation appears as insanity[1]; and in the second place, in
order to prove a causal connexion, in addition to this
it must be shown that insanity is actually more frequent in the descendants of drunkards than in those
of the sober. The undoubted evils of excessive drinking are many and obvious enough, but it does not
follow that physical or mental degeneration of the
descendants are among them, and it may be a false
hope to suppose that these evils could be removed
merely by the abolition of drink.

The same sort of argument may apply to the
undoubted physical and mental inferiority of our
slum population. It is not yet proved whether this
is the effect of miserable surroundings, or whether
the 'unfit' gravitate to the worst places because the
more fit occupy the better. These are problems
which society has as yet scarcely attempted to face,
and yet it is clear that on their correct solution
depends the central question of social reform. If
man is to any appreciable extent the creature of his
environment, then improved conditions will improve
the race. But if, as the study of heredity suggests,
though it would be rash to say it is proved, man is
almost entirely the product of inborn factors which

---

[1] In this connexion see Barrington and Pearson [10].

are little affected by environment, then improved
conditions may only encourage the propagation of
the degenerate, and the race as a whole may go back
rather than forward. Responsible students are not
lacking who maintain that this is already taking
place. It is said that the increase of insanity which
is believed to have taken place in modern times is
due to the provision of asylums where the insane
are properly cared for and frequently discharged as
'cured.' When the insane were treated on the 'strait
jacket' system no cure could be effected, and so the
unfortunates could not recover to propagate their
kind. But on the present system it not infrequently
happens that the insane are enabled to bring into the
world large families, so that it is not improbable that
the increase in number may be due to this, rather
than to the increased strain of modern conditions.
No one would advocate a return to the old system,
but some restriction on the reproduction of the mentally deficient is undoubtedly demanded by modern
knowledge of heredity.

It is even said that hospitals and the feeding
of destitute school-children are really working in the
direction opposite to what is intended, by enabling
the degenerate to live and beget families, who under
harder conditions would never have survived[1]. If a

[1] It is of course not suggested that all or even the majority of those
who receive such help are degenerate, but it can hardly be doubted
that a very high proportion of the 'unfit' will take advantage of it.

child is to survive it is undoubtedly better that he should be well fed and cared for, but looking at the matter apart from all sentiment, it is quite possible that posterity will be worse rather than better as a result of such institutions. It is not improbable that future generations will find that our methods for the relief of distress are on wrong lines, and that other means must be found for dealing with the problem, which will cure the evil at its root instead of attempting to alleviate the symptoms.

Another point at which the study of heredity touches social problems is the treatment of criminals. It is becoming recognised that a certain proportion of criminals are in some way abnormal, and that their crimes are due not to evil surroundings nor to wilful perversity, but to inherited defects. If this is actually the case, penal treatment of such is no less cruel than similar treatment of the insane, but in both cases efforts at reclamation or cure, followed by liberty and encouragement to marry, may only lead to a repetition of the same evils in the next generation. The present teaching of biology is perfectly clear, that in the case of the evils mentioned above and many others, marriage of those afflicted, and to a less extent of their near relatives, involves a grave risk of transmitting the affection to descendants, and so of inflicting serious injury upon society[1].

[1] See [43].

CHAPTER IX

In all the cases discussed hitherto the characters considered have been transmitted independently of the sex of the parents or offspring. The results of reciprocal matings have been identical, and the distribution of the characters among the offspring has been independent of their sex. It is well known, however, that many characters, in animals especially, are confined to one sex or are developed differently in males and females; this is most conspicuously so in 'secondary sexual characters,' that is to say, features characteristic of one sex only but not in any way directly connected with reproduction. As examples may be mentioned the distinctive plumage of many male birds, or patterns of butterflies, the horns of male mammals, and the hair on the face in men. Darwin long ago showed that such distinctively male characters may be transmitted through the female, and it has recently been shown that for Pheasants the converse is also true. Before entering in greater detail into the discussion of such cases it will be simpler to consider some other instances of 'sex-limited' or sex-modified inheritance which have been worked out in recent years, and which will probably throw light on sexual dimorphism in general.

It has been seen that in Mendelian inheritance one allelomorphic character is commonly dominant over the other, but that many cases are known in which dominance is partial, and it is probable that in some instances a positive or 'present' character may be recessive in the sense that it cannot appear unless introduced into the individual (zygote) from both parents. Now recent work has shown that some characters are dominant in one sex but recessive or only partially visible in the other (when heterozygous). One of the first cases of this to be discovered was the Tortoiseshell and Yellow Cat, and although there are probably complications in this instance which are not fully worked out, yet it is fairly certain that in cats which are in constitution heterozygotes between yellow and black, in males the yellow is completely dominant, but in females the dominance is partial, giving tortoiseshell. That is to say, in yellow-black heterozygotes, in males the yellow colour completely suppresses the black, but in females the black appears with the yellow. Thus tortoiseshell cats are almost invariably females, the corresponding males being yellow. A more completely known case is that of horns in Sheep. If a sheep of a breed in which both sexes bear horns is mated with one of a completely hornless breed, among the F_1 offspring the rams bear horns and the ewes are hornless. Such F_1 young bred together give in F_2 both horned and hornless males and females, and all the horned ewes and hornless rams are homozygous and breed true, while the hornless ewes and horned rams may be either pure or heterozygous. Here we have a clear case of a character which is dominant in one sex (male) and recessive in the other (female).

In such a case as this it may be said that the character is *transmitted* independently of sex, but its *appearance* in heterozygous offspring is modified or controlled by the sex of the individual. Both males and females receive the character alike, but it appears in males while it is suppressed in females of the same constitution.

In another important group of cases the actual distribution of the character among the offspring is determined by their sex, and a study of such throws important light upon the vexed question of sex-determination and the nature of sex itself. As an instance of this we will first consider one of the best-known examples, that of the inheritance of the variety *lacticolor* in the common Currant Moth (*Abraxas grossulariata*). The very rare variety *lacticolor* is found wild almost solely in females. When mated with a typical (*grossulariata*) male all the offspring of both sexes are typical, i.e. the type is completely dominant over the variety *lacticolor*. When two such heterozygous individuals are paired together, *lacticolor* appears as would be expected in about one quarter of the offspring, but *all the lacticolor specimens are females.* When a *lacticolor* female is mated with a heterozygous male, half the males and half the females

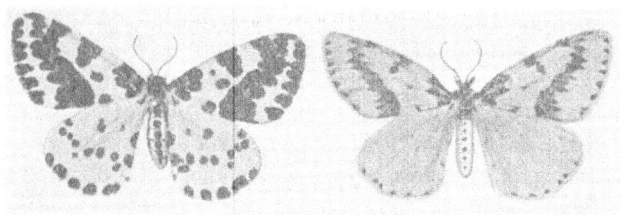

Fig. 13.   *Abraxas grossulariata* and its var. *lacticolor.*

are *lacticolor*, that is to say, by this mating alone can the *lacticolor* character be transferred to the male sex. *Lact.* male × *lact.* female gives only *lact.* in both sexes, as would be expected from two recessives paired together, but *lact.* male by heterozygous female gives *all the females lact., all the males typical (heterozygous) grossulariata.* Using the abbreviations *L* for variety *lacticolor*, *G* for the typical *grossulariata* character, *G* (*L*) for heterozygous *grossulariata*, ♂ for male,

and ♀ for female, these results may be summarised
thus:

1. L ♀ × G ♂   gives *gross.* ♂, *gross.* ♀.
2. $G(L)$ ♀ × $G(L)$ ♂ gives *gross.* ♂, *gross.* ♀, *lact.* ♀.
3. L ♀ × $G(L)$ ♂   gives *gross.* ♂, *lact.* ♂, *gross.* ♀, *lact.* ♀.
4. $G(L)$ ♀ × L ♂   gives *gross.* ♂, *lact.* ♀.
5. L ♀ × L ♂   gives *lact.* ♂, *lact.* ♀.

The important points about these facts are that
lacticolor is clearly a recessive variety when crossed
with *grossulariata*, but the distribution of the character among the sexes is different in the reciprocal
crosses, nos. 3 and 4 above. The only explanation
that has been offered to account for this is that
*females lay two kinds of eggs, one destined to become
males and the other to become females, and that the
female-bearing eggs cannot carry the grossulariata
factor.* Thus in mating no. 4 above (using G and L
for *gross.* and *lact.*), the heterozygous ($G(L)$) female
produces male eggs bearing G and female eggs bearing
L; all the spermatozoa (germ-cells) of the male bear
L, and thus the offspring are GL (= *gross.*) males
and LL (= *lact.*) females. But in the converse cross
no. 3, the *lact.* female (LL) produces male and female
eggs both bearing L, the heterozygous (GL) male
produces equal numbers of G and L spermatozoa,
and thus the offspring are GL and LL males, GL and
LL females. We thus arrive at the conclusion that

in this species at least the sex is determined by the egg of the female parent, even before fertilisation.

Perhaps the most remarkable fact in the whole case remains to be mentioned, that, as far as experiment has shown, pure (homozygous) females of the moth *A. grossulariata* do not exist; all the wild and apparently pure *gross.* females that have been tested by pairing with *lact.* males have given *lact.* female and *gross.* male offspring, so showing that they are in constitution *GL.* Since, however, wild males are pure *gross.* (*GG*) and since *GL* females produce *G*-bearing male eggs and *L*-bearing female eggs, the *lact.* (*L*) factor never normally becomes transferred to the male sex. Similar cases have been discovered in the inheritance of pink eyes in Canaries, and more recently for more than one pair of characters in Fowls. Of the latter, the most noteworthy is the barring of the Plymouth Rock, for which all barred hens are found to be heterozygous when mated with an unbarred (black) cock.

The essential feature of the explanation given above is that eggs of two kinds, male-producing and female-producing, are laid in equal numbers, and that the character for femaleness cannot be associated in the same egg with the *gross.* factor, so that female-bearing eggs must always bear *lact.* The case is probably comparable with the 'gametic coupling' which has been found between characters belonging to different allelomorphic pairs in the sweet-pea and other flowers, quite apart from sex. Sweet-pea flowers are of two shapes, with 'erect' standards or with 'hooded' standards. The erect character is dominant to the hooded. Among the offspring of certain sweet-peas heterozygous for this pair of characters and also for purple and red flowers, the purple offspring include both erect and hooded standards, but the reds are nearly all erect. This means that in the formation of the germ-cells the purple factor and the erect factor go into different germ-cells, so that instead of four kinds of germ-cells being produced in equal numbers, two are produced in great excess, viz. purple-hooded and red-erect. These meeting at random give purple-hooded, purple-erect and red-erect, but hardly any red-hooded.

It has been discovered by Bateson and Punnett that the purple and erect factors go chiefly into different germ-cells if introduced into the heterozygote by different parents, but are 'coupled' together and go into the same germ-cell if introduced from the same parent. In the case of the Currant Moth the G character is always introduced into the female from the male and possibly for this reason is never borne by female-bearing eggs.

It has been seen that the case of the Currant Moth and the similar cases in Fowls and Canaries lead to the inevitable conclusion that in these animals there are two kinds of eggs, male-bearing and female-bearing. In other cases, however, exactly similar evidence shows that the germ-cells of the male (spermatozoa) are of two kinds. The example of this which has been most fully worked out is the Fly *Drosophila*, in which a series of remarkable discoveries have been made in America by Morgan. He has found many sex-limited characters, perhaps the simplest of which is eye-colour. The normal flies have red eyes; a white-eyed male appeared and when mated with a red-eyed female gave only red-eyed offspring, both males and females. When, however, these were mated together, all the female offspring were red-eyed, but about half the males had white eyes. A white-eyed male mated with a heterozygous red-eyed female gave red and white eyes in both sexes, but any red-eyed male mated with a white-eyed female gave all the males white, all the females red-eyed. This case then is in every respect precisely the converse of what is found in the Currant Moth; the males must be heterozygous for a sex-determiner, and the red factor in the male is transmitted only to the female offspring.

Apparently similar conditions are found in Man in certain diseases of the eye, e.g. colour-blindness and certain forms of night-blindness, and in the disease known as Haemophilia. Here the affected male appears to transmit the factor for the disease only to his daughters, for sons of affected men are normal and do not transmit, while the daughters of affected men, though not themselves affected, transmit the disease to some of their sons.

The suggested explanation is that 'normality' is represented by a factor for which the female is homozygous (NN), the male heterozygous (Nn). If in the male the normality factor N is replaced by the factor for colour-blindness, for example (C), it will cause the appearance of the disease in the male for there is no second N to counteract C. The male transmits N or C only to his daughters, but a woman who receives C from her father will receive N from her mother, and so will not be colour-blind. She will transmit C, however, to half her children, so that some of her sons will be colour-blind. This is illustrated in the following scheme.

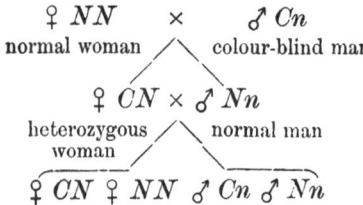

In the fly *Drosophila* and in Man, then, it is clear that the germ-cells of the male must be of two kinds, while we have seen that precisely similar evidence from a moth and two kinds of birds shows that in them the female produces two kinds of germ-cells.

The conclusion has been drawn by some students of the subject that in the Moth, Canary and Fowl the egg determines the sex of the offspring, while in Man and the fly *Drosophila* the spermatozoon determines the sex. There is some evidence, however, that *both* modes of inheritance may exist in Fowls, which would contradict this supposition, and for this and other reasons it is perhaps more likely that in animals generally, and probably in plants also, there are two kinds of egg-cells (germ-cells produced by the female) and two kinds of spermatozoa (germ-cells of the male) and that an egg-cell destined to produce a female is fertilised by a spermatozoon of one kind, a male-producing egg by a spermatozoon of the other kind. The facts in any case indicate that the determiners for maleness and femaleness are comparable with Mendelian characters, which segregate from each other in the formation of the germ-cells as do the 'factors' for Mendelian allelomorphs. For a more detailed presentation of this hypothesis and the evidence on which it rests, the reader must be referred to the works cited in the bibliography [especially 2, 9*a*, 22*b*, 43*a*].

It will be noticed that in the description of the nature of sex-determination here barely outlined, the sex of an individual is regarded as being no less irrevocably determined by the factors contained in the germ-cells than is the appearance of an inherited Mendelian character. This view is opposed to the older belief that the sex is determined by nutrition and other environmental conditions acting on the embryo or larva. It may be well therefore to summarise the evidence for the belief that sex is essentially inherited, and not determined by conditions. The evidence is of several distinct kinds. In the first place, in a variety of animals (and in Man in the case of 'identical twins'), whenever one egg divides to form two or more individuals, these are always of the same sex. Secondly, several animals belonging to widely diverse groups produce two kinds of eggs, one yielding males, the other females, and in these cases the sex is clearly determined before development begins. Thirdly, in the Bee and other forms, a fertilised egg yields females, an unfertilised egg, males; in this case fertilisation appears to determine the sex, but conditions operating later have no power to change it. In all these widely distributed cases we have direct evidence that sex is determined from the fertilisation of the egg or before it, presumably by 'sex-determinants' present in the germ-cells.

Evidence of a different nature has been afforded by the study of the development of the germ-cells, especially but not exclusively in Insects. In the nucleus of the developing cell in a number of species it is found that the bodies known as chromosomes (see Appendix, pp. 143, 148) have an even number in the female and *one less* in the male, for example in one case 22 in the female and 21 in the male. In the female two of these differ visibly from the rest, in the male there is one odd one, the remaining twenty being like the corresponding twenty of the female. At the 'maturation divisions,' when the chromosome number is halved, 11 go into each mature egg, but in the male the odd one goes into half the spermatozoa, leaving the other half without it. All eggs thus contain $10 + 1$, but spermatozoa are formed in equal numbers having $10 + 1$ and 10. In fertilisation then two kinds of zygotes will be produced, those made by $(10 + 1)$ egg $+$ $(10 + 1)$ spermatozoon giving $20 + 2$ (female), and those made by $(10 + 1) + 10$ giving $20 + 1$ (male). On the hypothesis of the nature of sex outlined above it is supposed that the odd chromosomes are related to the sex-determinants (if they are not themselves the determinants), so that the sex is determined by the combination of chromosomes in the nucleus. In any case, since the numbers are regularly found, external conditions can have no part in deciding the sex of the individual.

Finally, the cases of relation between sex and inherited characters described above (*lacticolor*, colour-blindness etc.) leave no room for the action of environment after the individual has once begun to develop. But it should be pointed out that environment may influence the *proportion* of male-producing and female-producing germ-cells which come to maturity, and a considerable amount of evidence has been collected showing that the varying proportions of the sexes under different conditions may be due to this cause. If some conditions favour the output of male-producing eggs, others of female-producing, the conditions will then indirectly affect the sex of the offspring, not by changing it in the individual, but by causing more individuals of one or the other sex to be born.

One of the reasons which have led biologists to assume that environment may modify the sex of the individual is the fact that changed conditions may influence the extent to which the sexual characters develop. Worker bees for example are females in which modified food and surroundings have prevented the full development of the female characters. The characters proper to one sex frequently are found in a rudimentary condition in the other, and in abnormal cases may develop. Or these characters may be prevented from developing in the sex which normally exhibits them by certain conditions, especially by removing the ovary or testis from the young animal, for example the non-appearance of horns in castrated deer or the disappearance of the distinctive sex-features in Crabs in which the reproductive organs

are destroyed by parasites[1]. Cases of this kind therefore bring us back to the question of 'secondary sexual
characters' and sexual dimorphism which were briefly
touched upon at the beginning of this chapter.

In discussing the determination of sex, the nature
of sex itself must be sharply distinguished from the
secondary characters which are normally associated
with it. The essential difference between the sexes
is the presence of testes in the male and ovaries in
the female ; the associated characters, which may or
may not be directly connected with reproduction, are
certainly sometimes and may perhaps always be
clearly separated into a distinct category. It has
been seen that the evidence is conclusive that sex
often, and probably always, is determined by the
germ-cells, i.e. from the moment of fertilisation if not
before. But it is equally clear that the characters
associated with one or the other sex are not determined in the same way ; their presence or the
extent to which they are developed may be to some
extent dependent upon environment. This is shown
by the case of the worker-bee ; by the effects of cas-
tration in preventing the appearance of distinctively
sexual characters ; by the appearance in old or

---

[1] In this particular instance it has been found that a male crab
affected by the parasite may even produce eggs, and it must be supposed
that femaleness is present in a recessive condition, so to speak, in the
male, and is caused to appear by the action of the parasite (*Sacculina*).

sterile female birds of plumage resembling that of
the male; and lastly by the occasional development
in one sex of characters which are normally rudimentary in that sex, but are well developed in the
other, such as the breasts in man. If such characters
are not directly connected with reproduction, for
example the splendid plumage of some male birds or
the horns of deer, they are called secondary sexual
characters, but there is no sharp distinction between
these and accessory sexual structures such as the egg-
laying apparatus of female insects.

It was said above that the appearance of sexual
characters is to some extent dependent on environment, but in fact the only changes which can modify
their development are those which affect the essential
reproductive organs, the ovaries and testes. When
these are removed or suppressed, the sexual characters
may be modified; their appearance indeed depends
in many cases upon the presence of functional sexual
organs, and frequently they do not appear until the
sexual organs are mature. Now it has been seen
above that certain characters are dominant in one sex
and recessive in the other, and the most typical case
of this is in an actual secondary sexual character—
horns of sheep, for when horns are present they are
always better developed in the male. Their 'recessive'
condition in a ewe produced by crossing a horned by
a hornless breed must probably be ascribed to a

factor, confined to the female, which prevents horns from developing even though the factor for horned-ness is present. Some such inhibiting factor, confined to one sex, may possibly explain sexual dimorphism in general. It may be of course that in some cases the inheritance of sexual characters is limited like that of the variety *lacticolor* in the Currant Moth ; if the var. *lacticolor* were dominant instead of recessive, we should have an instance of ordinary sexual dimorphism in this case, for all females are normally heterozygous and would therefore show one character, all males homozygous, and would show the other. Some condition of this kind will probably be found to apply to the remarkable cases of certain butterflies which have two or three discontinuous and very distinct forms of the female, but only one kind of male, and in which all the different forms have been bred from eggs laid by one female insect. The few breeding experiments that have yet been made with these species (Papilios or 'Swallow-tails' from Africa and India) suggest that a combination of alternative dominance and sex-limited inheritance will probably be found to explain them. It seems reasonable to suppose, therefore, that not only the determination of sex itself, but the difficult problems of sexual dimorphism and the inheritance of secondary sexual characters, will all be found on analysis to fall into line with the simple principles of Mendelian Heredity.

APPENDIX I

In the foregoing pages, Heredity has been regarded
as the relation between parents and offspring in
respect of their bodily characters, and it has been
shown that this relation has been brought about in
some way by the germ-cells, but very little has been
said about the mechanism by which this is effected.
This side of the question is very largely speculative,
and in order to keep speculation to some extent
apart from facts, an account of theories of hereditary
transmission, and of recent work on the supposed
material basis of hereditary characters, has been
postponed to appendices, which the reader who seeks
for facts and well-founded deductions alone, may
omit at will. First, a summary of the chief theories
of heredity will be given, and then a short account of
recent work on the subject.

In the introduction it was pointed out how closely
related are our ideas of Heredity and Variation with

theories of Evolution, and thus the history of the two
subjects largely coincides. The first important theory
was that of Lamarck, published in 1809, and although
it had little influence at the time, in more recent
years Lamarck's main principle has found many supporters. His theory was essentially that 'acquired'
modifications are being continually produced and
perfected by every organism during its life, and that
they are at least partially transmitted to its offspring,
so that each generation will be rather better adapted
to its surroundings than its predecessor. In this way,
for example, the great length of the neck of the
giraffe would be explained by the continual striving
through many generations to reach higher leaves on
the trees ; or the limbless condition of snakes and
slow-worms by the gradual loss of limbs through
disuse. But it has been seen that the assumption
that acquired characters are inherited is open to
grave doubt, and hence the followers of Lamarck
are fewer at the present time than formerly.

Darwin's great theory of Evolution by Natural
Selection of course depends on quite different principles, but it, like Lamarck's, is based essentially upon
the laws of variation and heredity. Darwin himself
made astonishing progress in the investigation of
these laws, and although he would doubtless have
been the first to admit the incompleteness of our
knowledge, yet he collected sufficient evidence to

enable him to formulate one of the first really important theories of heredity, which he called the Theory
of Pangenesis [7, (1868)]. The essence of his theory
was that every cell of an organism gives off minute
particles or 'Gemmules' from itself, which circulate
in the body and finally come to rest in the germ-cells,
or in parts where buds may be developed. The
gemmules were regarded as being capable of multi-
plication, and of transmission to a future generation
in a dormant state. They were supposed to be given
off from all tissues at every stage of development, so
that every unit of the organism at every stage would
be represented in the germ-cells. On the development of the germ-cell, the contained gemmules
would give rise to cells like those from which they
were derived, and so the characters of one generation
would be transmitted to those which follow.

By this hypothesis Darwin accounted for the
phenomena of sexual and non-sexual reproduction,
regeneration of lost parts, variability, inheritance
both of inborn and acquired characters, and lastly
of reversion to a previous ancestor. The hypothesis
was one of the first which attempted to bring all
these various groups of facts into line, but it had
the serious defect that there was no direct evidence
whatever for the existence of gemmules, and, assuming
their existence, to be accommodated in the germ-cells
they must be so exceedingly minute as to be almost

unimaginable. The Theory of Pangenesis never gained
any very wide acceptance, but is of great importance
owing to its stimulating effect on later work and
thought. To a great extent it led to the formulation
of other theories of heredity[1], any account of which is
prevented by limitation of space. It can only be
mentioned that the chief hypotheses which followed
Darwin's laid successively more and more emphasis
upon the idea that the germ-cells are not made up of
samples taken from the body, but have a certain
independence. So grew up the conception of 'germinal
continuity,' that is, the idea that the germ-cell of one
generation gives rise not only to the body of the next,
but also directly to its germ-cells, so that the body
does not produce germ-cells, but only contains them.
We must now turn to the theory in which this idea finds
its most celebrated expression, Weismann's *Theory
of the Germ-plasm* (1885) [40, 8].

It is impossible in a short space to give an
adequate account of Weismann's great theory, which
he has worked out in fuller minuteness of detail than
has been done with any other theory of heredity, and
by which he has done more to stimulate discussion
and research than perhaps any biologist since Darwin.

[1] For a summary of the more important theories of heredity,
especially those of Herbert Spencer (1863, i.e. before Darwin's theory
of Pangenesis), Galton (1875) and de Vries (1889), see Thomson's
Heredity [33]. See also [8].

Weismann was led by his work on the origin of the germ-cells to a belief in germinal continuity as explained above, but the facts of regeneration of lost parts and other related phenomena caused him to give up the idea that a sharp distinction could be drawn between the cells of the body and the germ, and to substitute for it the idea of a distinction between body-substance and germ-substance, or as he calls it, body-plasm and germ-plasm. According to this hypothesis, the egg contains germ-plasm derived from that of the parent, and as the egg develops the germ-plasm increases and becomes distributed among the cells, and gradually, as the cells become specialised to form the different parts, the germ-plasm becomes converted into body-plasm and builds up the varied kinds of cells of the body. But some cells continue to possess the full complement of ancestral germ-plasm, and these will go to form the germ-cells of the next generation. When an organ remains capable of regenerating lost parts, it is assumed that germ-plasm having the power to develop such parts remains in the cells and becomes active when required. Germ-plasm can thus be converted into body-plasm, but body-plasm cannot become germ-plasm, and hence Weismann assumes that no change brought about in the body (by environment, etc.) but not affecting the germ-cells, can be inherited by subsequent generations. It is therefore impossible according to his theory, that 'acquired characters' in the technical sense should ever be inherited. The germ-plasm of one generation gives origin to the germ-plasm of the next, and no external conditions acting on the body which contains and nourishes the germ-plasm can have effects which are transmitted unless the germ-plasm itself is altered.

Weismann in a series of books and papers has built up a detailed and highly complicated and speculative scheme of the nature and composition of the germ-plasm, only a brief summary of which can be given here. Much of it will doubtless not stand the test of fuller investigation, and parts of it are already discredited; but it has had the merit of stimulating an immense amount of valuable research, and there are indications that some of his fundamental ideas will form the foundations of a true theory of the material basis of heredity.

Weismann assumes that the germ-plasm is contained in the nucleus of the cell, and, in particular, in the bodies known as chromosomes. Every nucleus contains a number of these bodies, in the ordinary condition of the nucleus distributed through its substance so as to be unrecognisable, but when the cell is about to divide they make their appearance as rod-like bodies whose number in general is constant in the nuclei of the same species of animal or

plant. Before the nucleus divides the chromosomes
split longitudinally so that they are accurately halved
and the two halves of each go into different daughter-
nuclei. Weismann supposes that the germ-plasm
is contained in the chromosomes, and consists of
numerous units with different properties. When the
chromosome splits, each unit is supposed to divide
into two similar halves, and thus each daughter-
nucleus receives a similar complement of germ-plasm.

In the union of male and female cells in fertilisation, the nucleus of each cell brings its complement
of chromosomes, and thus if there were no special
provision, the number of chromosomes would be
doubled in each generation. But it is actually found
that in the cell-divisions immediately preceding the
development of both male and female sex-cells, a
process occurs which results in the removal of half
the chromosomes from the nucleus, and thus when
the male and female nuclei unite the normal number
of chromosomes for the species is restored. Since
Weismann regards the chromosomes as consisting
of germ-plasm, and as made up of a vast number
of units, each of which is the determinant for one
hereditary character, he saw that, without some such
process of removal of chromosomes in the formation
of the sex-cells, the germ-plasm must in a few generations become infinitely complicated. He therefore
predicted that some process of 'reduction' of chromo-

somes must occur, either by elimination of complete chromosomes or by transverse instead of longitudinal splitting, before any complete observations had been made showing that this actually happens.

Since Weismann supposes that the germ-plasm is contained in the chromosomes of the germ-cells, and since half the chromosomes are removed in the 'maturation' of these cells without preventing the transmission of any part to the offspring by inheritance, he concluded that each chromosome contains all the units ('determinants') necessary to a complete individual. (Later work has rendered this conclusion untenable: see Appendix II.) When fusion of male and female sex-cells takes place, the resulting individual will contain a mixture of the parental germ-plasms, the paternal in some chromosomes, the maternal in others. In the maturation of the sex-cells half these germ-plasms will be removed and in the next generation a fresh mixture will take place. It thus follows that the different chromosomes contain germ-plasms descended from different ancestors, and different mixtures of these will occur in different individuals. Here then we come to Weismann's hypothesis of the origin of variation. Since different individuals contain different combinations of ancestral germ-plasms, these will lead to varying effects in the development of the body; new combinations will be continually occurring in every

fertilised egg, and thus arises the variation between
separate individuals. Further, although by his theory
changes brought about in the body-plasm cannot be
transferred to the germ-plasm, yet influences acting
on the germ-plasm itself may modify it and so their
effects will be transmitted. The most important of
these influences is nourishment, which may favour
some units of the germ-plasm rather than others.
He further supposes that there may be competition
for nourishment among the different units ('determinants') so that some increase at the expense of
others, and if this process should be continued
through a series of generations, certain characters
would show a steady increase while others corre-
spondingly decrease. Variation thus arises by changes
brought about in the germ-plasm, and by the
recombination of varied ancestral germ-plasms in
each generation. Such variations will be inherited,
and in this respect will differ entirely from changes
brought about in the body during its life by the
action of environment.

It has been shown that in the earlier theories
of heredity it was assumed that the germ-cells were
produced by the body, and that they must therefore
be supposed either to contain samples of all parts
of it, or at least some kind of units derived from
those parts and able to cause their development
in the next generation. Gradually, as the study

of heredity and of the actual origin of the germ-cells has progressed, biologists have given up this view in favour of a belief in germinal continuity, that is, that the germ-substance is derived from previous germ-substance, the body being a kind of offshoot from it. The child is thus like its parent, not because it is produced from the parent, but because both child and parent are produced from the same stock of germ-plasm.

APPENDIX II

In Appendix I, it has been mentioned that Weismann regarded the chromosomes of the nucleus as the bearers of hereditary characters, and more recent work has provided striking confirmation of this belief. That some, if not all, the hereditary characters are determined by the *nucleus* of the germ-cell is indicated by several facts. In the first place, the spermatozoon consists of little else but a nucleus with a vibrating tail, and the tail may be shed as the spermatozoon enters the ovum. Secondly, experiments in fertilising non-nucleated fragments of sea-urchin eggs by sperm of a different species, give evidence that the hereditary characters of the resulting larvae are exclusively those of the paternal species. This conclusion however has been disputed, and can only be regarded as probable rather than certain. Again, experiments in fertilising one egg simultaneously by more than one spermatozoon, lead

Boveri [3] to believe that the subsequent development of the cells of the embryo depends on the distribution of the chromosomes in the abnormal divisions consequent on double fertilisation. And Herbst [16] has obtained sea-urchin larvae made by crossing distinct species, which on one side of the body resemble one parent, and on the other side the other parent. He shows that these differences are connected with differences in the size of the nuclei of the two sides, and that probably the part with maternal characters contains only maternal nuclear substance, while the part showing the paternal character has nuclei derived from both parents.

The foundation of the belief that the chromosomes bear the essential determinants for hereditary characters is provided by the behaviour of the chromosomes themselves in the maturation divisions of the germ-cells. It has been pointed out that at these nuclear divisions the chromosome number is halved, and restored to the full number again in the next generation by the union of two germ-cells each bearing the half-number. Now it has been found in certain cases that the chromosomes are not all alike, but differ among themselves in size and shape, and when this is so it can be seen that the nucleus just before maturation contains *two* of each kind. If the different kinds of chromosomes are represented by letters, A, B, C, D..., there will then be two A's, two B's, etc. in the nucleus. The actual processes in the reduction division are somewhat complex, but briefly they consist in a pairing together of the chromosomes, followed by a division of the nucleus in such a way that the two members of each pair are separated into different daughter-nuclei, so that the daughter-nuclei each contain half the full number. When the chromosomes differ among themselves, it is seen that two similar ones always pair together, i.e. A with A, B with B, etc. *Thus the daughter-nuclei each contain the whole series A, B, C..., but have only one of each, instead of two.*

If then it is imagined that each chromosome is the bearer of the determinant (or 'factor') for a Mendelian character, we may regard one individual as having a double series of chromosomes A, B, C..., etc., and another as bearing the allelomorphic characters a, b, c..., etc. When these individuals are mated, the heterozygote will bear both series, A and a, B and b, etc. In the formation of the germ-cells, A will segregate from a, B from b in exactly the way required by Mendelian theory. But there is no reason to suppose the series A, B, C... should all go into one germ-cell, and a, b, c... into the other; A may go into the first daughter-nucleus and a into the second, but b may go with A into the first, and B into the second. So in crossing races differing

in more than one allelomorphic pair, all possible combinations can be produced, except that no germ-cell
can contain both the members of one pair.

The suggestion that this segregation of chromosomes, which can be seen to take place, is the
mechanism by which the members of an allelomorphic
pair of characters are segregated, was at first sight
quite speculative; but it seemed exceedingly unlikely
that machinery so exactly adapted to bring it about
should be found in every developing germ-cell, if it
had no connexion with the segregation of characters
that is observed in experimental breeding. There
came the further fact in support of the suggestion, that
it is known in many insects that one pair of chromosomes is closely connected with sex, for in the males of
these species one chromosome of the pair is absent or
much reduced, but in the female both are similar.
These sex-chromosomes separate from one another
like the others (when both are present), and it has
been seen that there is experimental evidence for the
view that the sex-determiners behave like Mendelian
allelomorphs. One serious difficulty however suggests
itself at once; the chromosomes are limited in number,
and it is undoubted that more allelomorphic pairs of
characters may exist in a species than there are pairs
of chromosomes, although in such cases there is no
evidence that members of different pairs are always
associated together. Several suggestions have been

made to meet this difficulty, of which perhaps the
most adequate is that the chromosomes are not in-
divisible entities, but are composed of smaller units,
each of which corresponds with one Mendelian factor.
The chromosomes are not permanently present in
the distinct form which is seen during cell-division,
but during the resting condition of the nucleus their
substance becomes diffused over a network of threads,
only to be collected again into definite chromosomes,
having the same number and form as before, pre-
paratory to the next division. If each chromosome
consists of a series of units having a definite arrange-
ment, and these units become scattered in the
'resting phase,' but are re-collected in the same
order when the chromosomes are re-formed, it does
not seem unlikely that a unit N may take the place
of the corresponding unit n from the other chromosome of the pair, so that if the chromosome A
consisted at one division of units M, N, O..., and the
corresponding chromosome a consisted of m, n, o...,
after the resting stage N and n might have exchanged
places, and chromosome A would consist of M, n, O. .
and a of m, N, o.... By some process of this kind
it seems probable that the observed phenomena of
chromosome reduction would account for all the
facts of Mendelian segregation.

It must be stated quite clearly, however, that
the study of the possible relation between chromo-

somes and body-characters is as yet in its infancy ; and this brief note can only sketch the lines on which modern work seems to support Weismann's hypothesis that the chromosomes are the physical basis of inheritance. It will be seen that his suggestion that all the chromosomes are on the whole similar is not confirmed, but the evidence that chromosomes do bear factors for inherited characters is considerably stronger than when the idea was first put forward[1].

[1] The suggestion referred to in the note on p. 97 must also be borne in mind, that it is not the ' factor ' alone which determines the development of a character, but a physiological relation between the factor contained in a chromosome and the surrounding protoplasm. If the latter is altered from any cause, the relation may be changed and the character modified, just as the plants raised from one sample of seed may be modified by growing them in different soils. If, as is quite possible, the relation is a reciprocal one, the factor may in some cases be permanently modified, and there would thus be a mechanism for the transmission of acquired modifications. At present, however, it must be admitted that experiment hardly provides sufficient basis for speculations such as these.

LITERATURE LIST

*The Works marked * are general treatises suitable
for further study.*

1. Agar, W. E. Transmission of environmental effects from parent to offspring in *Simocephalus vetulus*. Phil. Trans. Roy. Soc. B. 203, 1913.

*2. Bateson, W Mendel's Principles of Heredity. 3rd Impression. Cambridge, 1913. (Full bibliography to date of publication.)

2a. —— Materials for the Study of Variation. London, 1894.

3. Boveri, E. Ergebnisse über die Konstitution der Chromatischen Substanz des Zellkerns. Jena, 1904.

4. Castle, W. E. Heredity of Hair-length in Guinea-pigs, and its Bearing on the Theory of Pure Gametes. Publ. Carnegie Inst. Washington, No. 49, 1906.

5. —— Studies of Inheritance in Rabbits. Publ. Carnegie Inst. No. 114, 1909.

*5a. Darbishire, A. D. Breeding and the Mendelian Discovery. Cassel, 1911.

*6. Darwin, C. The Origin of Species.

*7. —— Variation of Animals and Plants under Domestication.

*8. Darwin and Modern Science. Cambridge, 1909.

9. Davenport, C. B. Statistical Methods with Special Reference to Biological Variation. New York and London, 1899.

9a. Doncaster. L. The Determination of Sex. Cambridge, 1914.

10. Eugenics Laboratory. Publications of the Galton Laboratory for National Eugenics, University of London:— Especially, Elderton, E. M. Relative strength of Nature and Nurture. Heron, D. A First Study of the Statistics of Insanity. Schuster, E. Inheritance of Ability. Barrington, A. and Pearson, K. A Preliminary Study of Extreme Alcoholism in Adults.

11. Ewart, J. C. The Penycuik Experiments. London, 1899.

*12. Galton, F. Inquiries into Human Faculty and its Development. Macmillan, 1883. (Cheap Edition, J. M. Dent.)

*13. —— Natural Inheritance. Macmillan, 1889.

14. —— Essays in Eugenics. Eugenics Education Soc. London, 1909.

*15. —— Hereditary Genius. Macmillan, 1869.

16. Herbst, C. Vererbungsstudien, I—VI. Archiv für Entwicklungsmechanik, 1906—1909. Especially V—VI, 1907 and 1909.

17. Hurst, C. C. On the Inheritance of Eye-colour in Man. Proc. Roy. Soc. (B) Vol. 80. 1908.

18. Johannsen, W. Ueber Erblichkeit in Populationen und in Reinen Linien. Jena, 1903.

18a. Journal of Genetics, 1911—A quarterly journal containing important papers on Heredity and allied subjects, edited by Professors Bateson and Punnett. Camb. Univ. Press.

19. Kammerer, P. Vererbung erzwungener Fortpflanzungsanpassungen. Arch. f. Entwicklungsmechanik, Vols. XXV (1907) and XXVIII (1909).

*20. Kellogg, V. L. Darwinism Today. London and New York, 1907.

*21. Lock, R. H. Recent Progress in the Study of Variation, Heredity and Evolution. 5th Edition. Murray, 1920.

*22. Lotsy, J. P. Vorlesungen über Descendenztheorien. Jena, 1906.

22a. McDougall, D. T. Organic Response. American Naturalist. Vol. 45, p. 5, Jan. 1911.

22b. Morgan, T. H. The Physical Basis of Heredity. Lippincott, 1919, and The Mechanism of Mendelian Heredity, New York, 1915.

23. Nettleship, E. Bowman Lecture. Trans. Ophthalm. Soc. 1909.

23a. Pearl, R. The effect on the domestic food of the daily inhalation of ethyl alcohol. Journ. Exp. Zool. XXII. pp. 125, 165, 241.

*24. Pearson, K. The Grammar of Science. 3rd Edition, Part II. London, 1912.

25. —— Mathematical Contributions to the Theory of Evolution. Proc. and Trans. Roy. Soc. (A) 1896—1903; also numerous papers in Biometrika, 1902—, especially 'On the Laws of Inheritance in Man,' 1903, 1904; and 'The Law of Ancestral Heredity,' 1903.

26. —— Huxley Lecture of Anthropological Inst. of Gt. Britain and Ireland. Trans. Anthrop. Inst. 1903, p. 179.

*26a. Przibram, H. Experimental Zoologie. Vol. III. Phylogenese, 1910.

*27. Punnett, R. C. Mendelism. 5th Edition. Macmillan, 1919.

*28. Reid, Archdall. The Principles of Heredity. London, 1906.

*29. Romanes, J. G. Darwin and After Darwin. 3 vols. London, 1892—1897.

30. Royal Society. Reports to the Evolution Committee, I—V, 1902—1909.

31. Schuster, E. Hereditary Deafness. Biometrika, IV, 1906, p. 465.

31a. —— Eugenics. Collins, 1913.

32. Standfuss, M. Handbuch der Paläarktischen Grossschmetterlinge. Jena, 1896.

32a. Sumner, F. B. An Experimental Study of Somatic Modifications. Arch. f. Entwicklungsmechanik, Vol. XXX, Pt. II, p. 317.

32b. Stockard, C. R., and Papanicolau, G. A further study of the hereditary transmission of degeneracy and deformities by the descendants of alcoholized mammals, Amer. Nat. L. 1916, pp. 65, 144. Further studies on the modification of the germ-cells of mammals, Journ. Exp. Zool. xxvi, 1918, p. 119.

*33. Thomson, J. A. Heredity. Murray, 1908. (Very good bibliography.)

34. Tower, W. L. An Investigation of Evolution in Chrysomelid Beetles of the Genus *Leptinotarsa*. Publ. Carnegie Inst. Washington, No. 48, 1906.

35. Treasury of Human Inheritance. Publ. Galton Laboratory for Nat. Eugenics, London University.

*36. Vries, H. de. The Mutation Theory (Trans.). London, 1910.

*37. —— Species and Varieties, their origin through Mutation. Chicago and London, 1905.

*38. Wallace, A. R. Darwinism. London, 1889.

39. Weismann, A. Essays upon Heredity and Kindred Subjects (Trans.). Oxford, 1891, 1892.

*40. —— The Germ Plasm (Trans. Parker and Rönnfeldt). London, 1893.

*41. —— The Evolution Theory (Trans. J. A. and M. R. Thomson). London, 1904.

42. Wheldale, Muriel. Plant Oxydases and the Chemical Inter-relationships of Colour-varieties. Progressus Rei Botanicae, iii, p. 457. Jena, 1910.

*43. Whetham, W. C. D. and C. The Family and the Nation. London, 1910.

43a. Wilson, E. B. Studies on Chromosomes. Journal of Exp. Zoology. 1905—1909.

44. Yule, G. U. Mendel's Laws and their probable relations to intra-racial Heredity. New Phytologist. Vol. i, 1902, p. 193.

GLOSSARY

Acquired Character. A feature developed during the life of the individual possessing it, in response to the action of use or environment.

Albino. An animal without pigment in the skin, hair or eyes. The hair is white; the eyes pink owing to the colour of the blood. Among plants, white-flowered varieties are called albinos. The condition is called *albinism.*

Allelomorph. One of a pair of alternative Mendelian characters. When a pair of characters are alternative in their inheritance, and segregate from each other in the formation of the germ-cells of an individual which contains both, they are said to be *allelomorphic* with each other. See *Segregation.*

Anther. The part of the stamen in a flower which contains the pollen.

Chromogen. A colourless substance which when oxidised gives rise to a coloured body (pigment).

Chromosome. A body in the nucleus of a cell, which absorbs stains and becomes clearly visible during nuclear division, but becomes dispersed through the nucleus during the resting phase. During nuclear division each chromosome becomes accurately halved, so that in general all cells of each species of animal or plant contain an equal number of chromosomes.

Continuous. See *Variation.*

Determinant. The hypothetical unit in a germ-cell which determines the production of a particular character in the individual derived from that germ-cell. See *Factor.*

Deviation. The amount by which an individual differs from the mode in continuous variation.

Dimorphism. The condition in which a species exists in two distinct types or sharply separable varieties. When the two sexes differ thus, the condition is called *Sexual Dimorphism.*

Dominant. When two varieties, differing in one character, are crossed together, and all the offspring have the character borne by one parent, that character is dominant. Applied to one of a pair of Mendelian allelomorphs.

Egg-cell. The germ-cell produced by the female.

Epistatic. When one character A is superposed upon another B, so that A prevents or obscures the appearance of B, although they are not allelomorphic with each other, A is said to be *epistatic* to B.

'Extracted' Homozygote. When two heterozygous individuals are mated together, their homozygous offspring are spoken of as 'extracted' homozygotes.

F_1, F_2. The Symbol F_1 is used to indicate the offspring (first filial generation) of a mating between two differing individuals. The later generations (second, third filial, etc.) are represented by F_2, F_3, etc.

Factor. In Mendelian inheritance, the hereditary determinant (q.v.) of a particular character is spoken of as the *factor* for that character.

Ferment. A body which has the power of causing chemical action between substances which in its absence are inactive towards one another.

Fertilisation. The union of male and female germ-cells which precedes the development of a new individual. It consists essentially in the fusion of the nuclei of the germ-cells.

Gamete. A germ-cell, q.v.

Gemmules. Hypothetical bodies supposed to be given off by the cells of the body, and entering the germ-cells, to transmit heritable characters to the next generation.

Germ-cell. A reproductive cell, which, usually after union with a germ-cell from another individual (fertilisation), develops into a new individual. In animals the germ-cells of the male are *spermatozoa*, those of the female *ova* (*egg-cells*). In plants the male germ-cells are contained in the *pollen* ; the female, *egg-cells*, in the ovules or embryo-seeds.

Germ-plasm. The germinal substance, which according to Weismann is alone able to give origin to new individuals.

Heterozygote. An individual containing both members of an allelomorphic pair of characters, i.e. which is hybrid in respect of that pair of characters, and produces germ-cells bearing one and the other respectively. Adjective—*heterozygous.*

Homozygote. An individual made by union of two germ-cells each of which bears the same member of an allelomorphic pair of characters, so that it is 'pure' in respect of that character, and all its germ-cells bear the same character. Adjective—*homozygous.*

Mode. The most frequent condition of a character which varies continuously. Its measurement is called the *modal value.*

Mutation. A variety which is not connected with the type by intermediates. More strictly, the sudden origin of such a variety.

Nucleus. A sharply defined body found in every cell, which seems to control the activities of the cell.

Ovum. The germ-cell produced by the female, an egg-cell.

Pin-eye. In *Primula* (Primrose, Cowslip, etc.), the form in which the style is long and the anthers low down in the flower-tube. The other form, with short style and anthers high up, is called *Thrum-eye.*

Pollen. The powder bearing the male germ-cells in a flowering plant.

Polymorphism. The condition in which a species exists in several distinct forms or varieties.

Recessive. When two individuals are crossed, bearing different members of an allelomorphic pair of characters, the member of the pair which does not appear in the offspring is called *recessive*.

Reversion. A 'throw-back' to a previous ancestor, or to the type of the species, when varieties are crossed.

Segregation. In Mendelian inheritance, the separation of the two characters of an allelomorphic pair, in a heterozygote, into distinct germ-cells, i.e. the formation of gametes each bearing one character of a pair, by an individual which contains both members.

Self-fertilisation. The fertilisation of a female gamete by a male gamete produced from the same individual. The process in plants is spoken of shortly as *selfing*.

Somatic. Having reference to the body ('*soma*') considered as distinct from the germ-cells. A character borne or exhibited by the body but not represented in the germ-cells is called *somatic* as contrasted with *germinal*.

Spermatozoon. See *Germ-Cell*.

Style. The part of a flower which receives the pollen, and conducts the male germ-cell to the egg-cell.

Telegony. The supposed influence of a former sire upon young born to a later sire by the same mother.

Type. The normal form of a species, which is regarded as *typical*.

Variation, Variability. The differing among themselves of individuals of the same species. When the extreme forms are connected by a complete series of intermediates, the variation is *Continuous*; when distinct forms occur, not connected by intermediates, it is *Discontinuous*.

Zygote. An individual produced by the union of two gametes.

INDEX

Ability, Inheritance of, 48
Abraxas, 124
Acquired Characters, 19, 24, 30, 45, 90–100, 139, 143
AGAR, 97
Albinism, 66, 71
Allelomorph, 56, 61, 71, 150
Amphidasys betularia, 88
Ancestral Heredity, Law of, 42
Andalusian Fowl, 69
BATESON, W., 19, 70, 76, 82, 128
Bee, 93, 132
BIFFEN, R. H., 64
Biometric Methods, 9, 32-51
BOVERI, T., 149
Brachydactyly, 108
BROWN-SÉQUARD, 95
Canary, 128
Cat, 123
Cataract, 109
Chromosomes, 133, 143, 148–150
Colour-blindness, 129, 130
Colour-inheritance, 71–83
Combs of Fowls, 66, 67
Correlation, 36
       ,,   parental, 39, 106
CUNNINGHAM, J. T., 91
Currant Moth, 124
DARWIN, 2, 8, 23, 24, 88, 139
DAVENPORT, 113
Deafness, 106
Dihybridism, 62
Disease, Inheritance of, 105
Dominant Characters, 54
Drosophila, 100, 129–131
Earwig, 17
ELDERTON, E. M., 118
Environment, 24–29, 117–121

Epistatic, 73, 75
Eugenics, 51
Evolution, 2, 88, 139
EWART, J. C., 101
Extracted Homozygote, 62, 86
Eye-colour, 66, 107, 112
Feeble-minded, 50
FISCHER, E., 27, 93
Flatfish, 91
Flower-colour, 66, 74, 77, 79–83
Fluctuation, 22, 45
Fowls, 66, 67, 99, 128
GALTON, Sir F., 24, 30, 33, 41, 117
Gametes, 57 (*see* Germ-cells)
Gametic coupling, 128
Gemmules, 140
Germ-cells, 57, 140,
Germinal Continuity, 142
Germ-plasm, 20, 23–26, 141–146
GREGORY, R. P., 82
Guinea-pig, 86, 94, 98
Habsburg lip, 113 (note)
Haemophilia, 129
Hair-colour, 66, 112
Hair-length, 66, 86
HERBST, C., 149
HERON, D., 106
Heterozygote, 59
Homostyle (*Primula*), 82
Homozygote, 61
Horns of Sheep, 84, 124
HURST, C. C., 107, 112
Hybridisation, 53
Insanity, 105, 118
Instinct, 92
JOHANNSEN, W., 44, 87
KAMMERER, P., 95
LAMARCK, 139

Leptinotarsa, 28
Lychnis, 66
Maize, 65, 66
Man, Heredity in, 104–121, 129
Marsh-Marigold, 15
Maternal Impression, 102
MᶜDougall, 96
Mendelian Heredity, 52–89
    ,,   ,,   in Man, 105, 107–115
Mendel's Experiments, 53, 54
Mental and Moral Characters, 49
Mid-parent, 40
Mode, 10, 34
Morgan, T. H., 100, 129
Mouse, 74
Mulatto, 113
Multiple factors, 88
Mutation, 22
Natural Selection, 8, 89, 139
Nettleship, E., 109, 111
Night-blindness, 109, 129
Nucleus, 143, 148
Organic Stability, 24
Pangenesis, 140
Pearl, 99, 118
Pearson, K., 32, 41, 119 (note)
Peas, 54
Pigeons, 66
Primula, 66, 79, 81–83
Protoplasm, 3
Punnett, R. C., 70, 128
'Pure Lines,' 44, 87
Rabbit, 74, 79, 87
Rat, 72, 79, 96
Recessive Characters, 56
Regression, 35
Retinitis pigmentosa, 111
Reversion, 74, 76
Rust in Wheat, 64
Sacculina, 135 (note)

Schuster, E., 49, 106
Sea-Urchin eggs, 148, 149
Seed-colour, 66
Segregation, 58
Sex, 84, 109, 122–137
Simocephalus, 97
Sheep, 84, 124
Silkworms, 94
Skew Curves, 15
    ,,   Correlation, 39
Skin-Colour, 113
Social Problems, 117
Standfuss, M., 27
Statistical Study of Variation, 32
Stature, 10, 34, 37
Sterility, 66, 115
Stockard, 98, 118
Stocks, 66, 80
Sumner, F. B., 96
Sweet-pea, 54, 66, 76, 128
Telegony, 101
Temperature, effects of, 25, 27, 28
Tortoiseshell Cat, 84, 123
Tower, W. L., 28, 96
Tuberculosis, 25
Twins, 30, 117, 132
Variation, 2, 5, 7
    ,,   Causes of, 8, 23–31, 145
Variation, Continuous and Discontinuous, 9–12, 17–19, 22, 88
Variation, Curves illustrating, 11–17
Variation induced by crossing, 29
Vries, H. de, 19
Wallace, A. R., 8
Weismann, A., 20, 23, 141–146
Weldon, W. F. R., 33
Wheat, 64, 66
Wheldale, Muriel, 77
Yule, G. U., 42
Zygote, 58